Data Science for Water Utilities

This addition to the Data Science Series introduces the principles of data science and the R language to the singular needs of water professionals. The book provides unique data and examples relevant to managing water utility and is sourced from the author's extensive experience.

Data Science for Water Utilities: Data as a Source of Value is an applied, practical guide that shows water professionals how to use data science to solve urban water management problems. Content develops through four case studies. The first looks at analysing water quality to ensure public health. The second considers customer feedback. The third case study introduces smart meter data. The guide flows easily from basic principles through code that, with each case study, increases in complexity. The last case study analyses data using basic machine learning.

Readers will be familiar with analysing data but do not need coding experience to use this book. The title will be essential reading for anyone seeking a practical introduction to data science and creating value with R.

CHAPMAN & HALL/CRC DATA SCIENCE SERIES

Reflecting the interdisciplinary nature of the field, this book series brings together researchers, practitioners, and instructors from statistics, computer science, machine learning, and analytics. The series will publish cutting-edge research, industry applications, and textbooks in data science.

The inclusion of concrete examples, applications, and methods is highly encouraged. The scope of the series includes titles in the areas of machine learning, pattern recognition, predictive analytics, business analytics, Big Data, visualization, programming, software, learning analytics, data wrangling, interactive graphics, and reproducible research.

Published Titles

Massive Graph Analytics
Edited by David Bader

Data Science
An Introduction
Tiffany-Anne Timbers, Trevor Campbell and Melissa Lee

Tree-Based Methods
A Practical Introduction with Applications in R
Brandon M. Greenwell

Urban Informatics
Using Big Data to Understand and Serve Communities
Daniel T. O'Brien

Introduction to Environmental Data Science
Jerry Douglas Davis

Hands-On Data Science for Librarians
Sarah Lin and Dorris Scott

Geographic Data Science with R
Visualizing and Analyzing Environmental Change
Michael C. Wimberly

Practitioner's Guide to Data Science
Hui Lin and Ming Li

Data Science and Analytics Strategy
An Emergent Design Approach
Kailash Awati and Alexander Scriven

Telling Stories with Data
With Applications in R
Rohan Alexander

Data Science for Sensory and Consumer Scientists
Thierry Worch, Julien Delarue, Vanessa Rios De Souza and John Ennis

Big Data Analytics
A Guide to Data Science Practitioners Making the Transition to Big Data
Ulrich Matter

Data Science in Practice
Tom Alby

Natural Language Processing in the Real World
Jyotika Singh

For more information about this series, please visit: https://www.routledge.com/Chapman--HallCRC-Data-Science-Series/book-series/CHDSS

Data Science for Water Utilities
Data as a Source of Value

Peter Prevos

CRC Press
Taylor & Francis Group
Boca Raton London New York

CRC Press is an imprint of the
Taylor & Francis Group, an **informa** business

A CHAPMAN & HALL BOOK

Designed cover image: Getty

First edition published 2023
by CRC Press
6000 Broken Sound Parkway NW, Suite 300, Boca Raton, FL 33487-2742

and by CRC Press
4 Park Square, Milton Park, Abingdon, Oxon, OX14 4RN

CRC Press is an imprint of Taylor & Francis Group, LLC

ISBN: 978-1-032-35455-2 (hbk)
ISBN: 978-1-032-35454-5 (pbk)
ISBN: 978-1-003-32697-7 (ebk)

DOI: 10.1201/9781003326977

Typeset in Alegreya Regular font
by KnowledgeWorks Global Ltd.

Publisher's note: This book has been prepared from camera-ready copy provided by the authors.

Contents

Preface

There were no computers in my first engineering job. Instead, we undertook all analysis using pencil, paper, and a desktop calculator. In the early 1990s, I managed projects in Europe and Asia using filing cabinets tightly packed with paper files. My assistant patiently typed this data into spreadsheets, which we used to analyse progress and plan future activities. Over the years, I must have created thousands of interconnected spreadsheets to create insights from this information.

While spreadsheets are flexible tools that combine data, analysis, and results, it quickly became apparent that managing a jungle of spreadsheets was arduous, to say the least. In addition, copying results between spreadsheets and written reports caused data integrity problems and copious errors and rework.

A former manager once warned me that my data analysis skills could become a curse. He suggested that my career would be limited to crunching numbers. I heeded his warning and focussed on managing projects rather than being a data analyst. But the world has changed since he gave me this advice. Twenty-five years later, my colleague Jenny Fogarty suggested "looking into this new thing called data science".

When I started writing my dissertation about customer service in water utilities, I quickly realised that spreadsheets would not give me the flexibility to manage such a complex project. So I started developing an interest in data science. As an avid free software user, I stumbled upon the R language, now my weapon of choice in solving analytical problems. These new skills changed my career as I now manage the data team at Coliban Water, a water utility in regional Australia.

The most powerful way to perfect a skill is to teach it. So, since 2019 I have been teaching the R language to water professionals in Australia. The syllabus for this course and the articles on my *The Devil is in the Data* website gradually grew into the foundations of the book you are now reading.[1]

The advice from my former manager is no longer relevant. Data scientists are the alchemists of business, converting data into golden insights. Unfortunately, there is much hype around data science, and some organisations need help to create value from their efforts. The main reason for these problems is a gap between the skills of subject-matter experts and data scientists. Some data science websites call people with both skills 'data science unicorns' as they don't appear to exist.

My motivation for writing this book is to breed data science unicorns by introducing water professionals to using code to solve problems. Teaching subject-matter experts to apply computer science to analyse data enables organisations to wield the incredible potential of advanced analytics.

This book came to completion with the support of the wonderful people of Water Research Australia. Karen Rouse, Kelly Hill, Carolyn Bellamy, and Jo Ohlmeyer have enabled me to teach water utility professionals the principles of data science and the R language. The syllabus for that course was the starting point for this book. The feedback from the participants of this course has helped to craft a program that teaches data science gradually and with specific reference to water management.

Khalil Mokhtari from the University of Abbes Laghrour in Khenchela, Algeria, helped me draw some diagrams using the PGF/TikZ language. He has helped me deliver the book on time and was generous in helping me understand this powerful graphical language better.

Peter Prevos, Kangaroo Flat, August 2022

---

[1]The Devil is in the Data, lucidmanager.org.

Foreword

Growing urban populations, ageing infrastructure, rising customer expectations, limited budgets, and the impact of climate change are increasingly putting a strain on water utility management. The Internet of Things (IoT) and the need to process incredible amounts of data from multiple, often disparate sources are essential to address such challenges. However, these solutions often generate a stream of *Big Data*, which is characterised by high volumes, varieties, velocities, and veracities of information, which is challenging to process and understand using traditional data management techniques. As a result, despite the relative maturity of big data technology and adoption in many other sectors, uptake within the water sector remains limited.

Today, *smart water* or data-driven technologies are changing how customers decide about their water use and how cities monitor and control their networks. A *smart water network* links multiple systems within a network to share data across platforms. This allows cities to better anticipate and react to water network issues, from detecting leaks and water quality incidents to conserving energy and tracking residential water consumption. By monitoring real-time information, network operators can stay informed about what is happening in the field and respond quickly and appropriately when a problem arises. This results in the city becoming more efficient and reducing the overall cost of service for the customer. However, the success of smart water projects often relies less on the actual technologies and more on the proper uptake by utility staff integrating these new solutions with their existing processes.

Before embarking on any smart water project, management needs to understand what data is right for them and develop a strategy which maximises its use. While global water utilities have long operated Supervisory Control and Data Acquisition (SCADA) systems and Geographic Information Systems (GIS) to monitor critical functions, the use of Big Data analytics, Artificial Intelligence (AI), and Machine Learning are now becoming more widespread. Interpreting such complex data often requires employing data scientists within utility staff or, more commonly, outsourcing these skillsets to outside consultants or technology providers through *as-a-service* solutions.

Big data management poses several challenges for water utilities, such as malfunctioning technologies, data errors, data storage needs, potential cybersecurity risks, data ownership concerns, and the lack of general standards and protocols. Within a water utility, there is often a short tolerance for new solutions, so data needs to be as accurate as possible to be entirely accepted. However, obtaining high data quality doesn't happen overnight. It is a constant learning process. While most water utilities still prefer to use Microsoft Excel to analyse data, there are far more effective methods such as statistical analysis, interactive dashboards, and integrated platforms.

To make any data management plan successful, there are a few critical lessons gained through the experience of the Smart Water Networks (SWAN) Forum. First, it is important to accept that some project steps may fail, but this is part of any innovation process, so the real question is what to do when one meets a hurdle – stop the project entirely or pivot to find alternative solutions? Second, data shouldn't be limited to a few people or withheld in internal silos; it is more valuable when widely shared across different utility departments and even outside the utility with customers and regulators. Third, it is not enough for data analysts to translate data. They must also be skilled communicators or *data storytellers* to convey their ideas effectively. Lastly, perhaps most important, utilities need to embrace new, modern ways of doing things to attract and maintain young talent.

It is vital to emphasise the human element in smart water, from how operators interact and engage with proposed smart water solutions through to increased data transparency made more accessible to customers. Smart water is a continuous and iterative process to reach a desired, strategic outcome to address the main business driver. The core of this journey is value creation, which refers to people, processes, and technologies that deliver smart water value across the organisation. This is essentially the 'why' behind smart water.

To truly appreciate this value, water professionals (in whatever organisation they work for) should learn the key principles of data science and how to apply them. This book provides the foundational knowledge for anyone working in the water industry to learn data science skills.

Amir Cahn, Executive Director, Smart Water Network (SWAN) Forum

1

Introduction

Water is an essential condition for life, but it is more than just a means to survive. It also plays a vital role in our culture and social interactions. People in the developed world enjoy a virtually unlimited supply of clean water merely by turning a tap. However, millions of people worldwide lack access to safe drinking water and can spend hours obtaining their daily needs. Managing precious water availability will become increasingly critical in a fast-warming world. The United Nations have defined seventeen *Sustainable Development Goals*.[1] Goal six calls for "availability and sustainable management of water and sanitation for all." Ensuring that all of humanity has access to a safe and reliable source of drinking water is thus one of the significant challenges of the twenty-first century.

Managing reliable water services requires a sufficient volume of water but also significant volumes of data. Water professionals continuously measure the flow and quality of water and assess how customers perceive their service. Professionals in the water industry rarely directly interact with the water they provide or the consumers they serve. Water utilities provide water and sewage services at 'arm's length' from the customers, so data is a proxy for the customer's experience.

Water data science, also called hydroinformatics, is an emerging discipline in water management. Water and data have a lot in common. Data professionals use data pipes and data lakes. They filter and clean data like a water engineer filters and transports water from a reservoir to consumers through physical pipes. Data and water are, as such, natural partners.

Traditionally, water professionals use a lot of spreadsheets to analyse data. While these versatile tools conveniently combine the data, analysis and results, they are unsuitable for complex applications. Many competing business intelligence tools, such as Power BI and Tableau, are available, making it easy to connect data to a visualisation and analysis tool. However, while these tools are excellent for presenting the analysis results, they don't give the user the same power over the data as a spreadsheet.

Data scientists analyse data with computer code, using languages such as Python, Julia, SQL, MATLAB, or, last but not least, R. Using computer code to analyse data provides unique benefits that neither a spreadsheet nor a business intelligence tool can provide. Using computer code is like writing a recipe to analyse data, which makes it easy to understand how the analysis works, preventing a black box that hides its workings. A computer language also has much more flexibility than clicking buttons on a screen. But with this great power comes great responsibility. Learning to analyse data with computer code does have a steep learning curve, mastering a plethora of new concepts and vocabulary. While this learning curve might seem daunting, keep in mind that:

The steeper the learning curve, the higher to payoff.

1.1 Who Is This Book For?

This book's content started as a course syllabus to teach water professionals how to use the R language. While the case studies are specific to the types of problems faced by water

---

[1]United Nations, *Sustainable Development Goals*. sdgs.un.org/goals

1.2.4   Case Study 4: Predicting Concrete Strength

The last case study uses data from a machine learning repository that describes concrete mixtures. Concrete is a versatile material that is mixed from various components and commonly used in water management assets, such as filtration plants and tanks. The amount of water, cement, sand, gravel, additives, and the curing time determine the compressive strength.

This data is used to explain the principles of machine learning, by predicting concrete strength using multiple linear regression and a decision tree.

1.2.5   Closing Chapter

The final chapter in this book consolidates the lessons from this book and provides some advice on how readers can best continue to learn R and continue their quest to improve their data science chops.

1.3   What Is Data Science?

The idea that data helps us understand the world is thus almost as old as humanity itself. It has gradually evolved into what we now call data science. Using data in organisations is also called business analytics or evidence-based management. There are also specific approaches, such as Six Sigma, that use statistical analysis to improve business processes. Although data science is merely a new term for something that has existed for decades, some recent developments have created a watershed between the old and new ways of analysing a business. The difference between traditional business analysis and the new world of data science is threefold (Prevos, 2019):

1. Businesses have much more data available than ever before
2. Increased computing power
3. Open-source innovation

This revolution is not only about powerful machine learning algorithms, but about a more scientific way of solving problems. Most analytical issues in supplying water or sewerage services do not require machine learning to solve. However, the definition of data science is not restricted to machine learning, big data, and artificial intelligence. These developments are essential aspects of data science but do not define the field.

One factor that spearheaded data science into popularity is the available toolkit, which has grown substantially in the past ten years. Open-source computing languages such as R and Python can implement complex algorithms previously in the domain of specialised software and supercomputers. As a result, Open-source software has accelerated innovation in how we analyse data and has placed complex machine learning within reach of anyone willing to make an effort to learn the skills.

Data science is a systematic and strategic approach to using data, mathematics, and computers to solve practical problems. A data scientist differs from a scientist who understands the world in generalised terms. The challenges facing data scientists are functional rather than scientific. A data scientist in an organisation is less interested in a generalised solution to a problem but focuses on improving how the organisation achieves its goals. In this sense, a data scientist is not strictly a scientist.

These three competencies are not the only qualities of a good data scientist. Other required skills are business competencies, such as writing a compelling business case, presentation skills,

1

Introduction

Water is an essential condition for life, but it is more than just a means to survive. It also plays a vital role in our culture and social interactions. People in the developed world enjoy a virtually unlimited supply of clean water merely by turning a tap. However, millions of people worldwide lack access to safe drinking water and can spend hours obtaining their daily needs. Managing precious water availability will become increasingly critical in a fast-warming world. The United Nations have defined seventeen *Sustainable Development Goals*.[1] Goal six calls for "availability and sustainable management of water and sanitation for all." Ensuring that all of humanity has access to a safe and reliable source of drinking water is thus one of the significant challenges of the twenty-first century.

Managing reliable water services requires a sufficient volume of water but also significant volumes of data. Water professionals continuously measure the flow and quality of water and assess how customers perceive their service. Professionals in the water industry rarely directly interact with the water they provide or the consumers they serve. Water utilities provide water and sewage services at 'arm's length' from the customers, so data is a proxy for the customer's experience.

Water data science, also called hydroinformatics, is an emerging discipline in water management. Water and data have a lot in common. Data professionals use data pipes and data lakes. They filter and clean data like a water engineer filters and transports water from a reservoir to consumers through physical pipes. Data and water are, as such, natural partners.

Traditionally, water professionals use a lot of spreadsheets to analyse data. While these versatile tools conveniently combine the data, analysis and results, they are unsuitable for complex applications. Many competing business intelligence tools, such as Power BI and Tableau, are available, making it easy to connect data to a visualisation and analysis tool. However, while these tools are excellent for presenting the analysis results, they don't give the user the same power over the data as a spreadsheet.

Data scientists analyse data with computer code, using languages such as Python, Julia, SQL, MATLAB, or, last but not least, R. Using computer code to analyse data provides unique benefits that neither a spreadsheet nor a business intelligence tool can provide. Using computer code is like writing a recipe to analyse data, which makes it easy to understand how the analysis works, preventing a black box that hides its workings. A computer language also has much more flexibility than clicking buttons on a screen. But with this great power comes great responsibility. Learning to analyse data with computer code does have a steep learning curve, mastering a plethora of new concepts and vocabulary. While this learning curve might seem daunting, keep in mind that:

The steeper the learning curve, the higher to payoff.

1.1 Who Is This Book For?

This book's content started as a course syllabus to teach water professionals how to use the R language. While the case studies are specific to the types of problems faced by water

---

[1]United Nations, *Sustainable Development Goals*. sdgs.un.org/goals

utilities, the principles of solving problems with code apply to anyone wanting to improve their skills in data analysis. This book is thus helpful for anyone interested to learn how to use the R language to systematically analyse data.

This book does not only show how to write code but also how to apply best-practice principles of data analysis and visualisation. Learning how to code is pretty straightforward, but using it to create water management outcomes requires additional skills. This book, therefore, provides a framework to produce *sound*, *useful*, and *aesthetic* data products. This chapter discusses the principles of this framework, which are applied with R code in the remainder of the book.

R is a popular programming language specially designed for data analysis with built-in statistical capabilities. The internet is awash with discussions on whether R or Python is better for data science. These discussions are like arguing whether Dutch or English are better for writing books. All languages, either natural or artificial, have their strengths and weaknesses. This book is based on principles, and the skills taught in these chapters are easily transferrable to Python or other programming languages.

1.1.1 Prerequisites

To benefit from this book, you must have some prior knowledge and experience with analysing data and statistics, either spreadsheets or otherwise. Experience with writing computer code is helpful but not necessary, as the book starts from the basic principles. Likewise, knowledge of water management is not required as the context of the case studies is explained in sufficient detail. Some knowledge of statistics is also helpful. Lastly, you will need access to the R language for statistical computing and the RStudio interface. Chapter 2 explains how to obtain and install this software.

1.2 Book Structure

The main objective of this book is to teach water professionals how to develop data science code to solve urban water management problems. The learning objectives for this book are:

1. Apply the principles of strategic data science to solve water problems
2. Understand the principles of writing sound code
3. Write R code to load, transform, analyse, and visualise data
4. Develop presentations, reports, and applications to share results

This book uses a case study approach to teach data science and uses R code to analyse data. Most other coding books focus on abstract examples before solving real-world problems. This approach is like learning a new human language by first learning grammar. Children learn playfully by using language and develop grammar skills as they gain new skills. This book uses the same playful attitude to learning to write data science code. Playing with the R language is more productive than aiming to understand all its theoretical foundations before using it in practice.

The R language contains an extensive collection of built-in data sets used for teaching. These examples are helpful but need more context. The case studies in this book use realistic problems that a typical water professional encounters. The data in the case studies is mostly fictitious, simulated using realistic assumptions in the fictional city of Gormsey.

Each chapter opens with learning objectives to guide the reader in developing these skills. This book should be read with RStudio open and reproducing the code. Unfortunately, no single book on data science can ever be complete. As such, each chapter also provides suggestions for further in-depth study.

Chapter 2 shows how to obtain and install R and RStudio, introduces the basics of the R language, and concludes with a mini-case study to practice basic arithmetic and assigning variables. The remainder of the book follows four case studies. Each case study starts with a problem statement and introduces readers to the relevant aspects of the R language. Readers then load, transform, explore and analyse the relevant data to solve the case study problem. The data and code for this book are available through the GitHub website (`https://github.com/pprevos/r4h2o`). Section 2.3.1 explains how to obtain this information.

To get the most from this book, make sure you understand what the code does, think about what the maths mean and comprehend the practical applications. The best way to achieve this goal is to copy the examples from the book and play with the code. The best way to learn computer code or anything else for that matter, is to play.

1.2.1 Case Study 1: Exploring Water Quality Data

The primary role of a water utility is to provide water that is safe to drink, which means that the water should be free from pathogens and chemical pollutants. One of the methods to ensure this is the case is by regularly sampling water and testing it in a laboratory.

This first case study discusses how to explore and visualise laboratory testing data from a supply network and present descriptive statistics. The case study revolves around checking the data for compliance with water quality regulations to minimise risk to human health. The case study ends with code that produces a PowerPoint presentation linked to water quality data.

1.2.2 Case Study 2: Understanding the Customer Experience

Water management is not only about cubic metres of water and milligrams of chemicals. Water professionals also need to know how to understand the voice of the customer. The second case study discusses how to collect and analyse customer survey data.

Water utilities are becoming ever more aware of their role in the community. Water professionals now also analyse the information they collect from customers. Data collected from living human beings requires a different approach to data gathered from scientific instruments. Measurement in the social sciences follows a different approach to measuring physical processes.

The second case study discusses how to clean data using code. This clean data is then analysed using factor analysis to ascertain the voice of the customer that emerges from the data. The second case study also uses cluster analysis to find segments of customers and linear regression to find relationships between survey questions.

1.2.3 Case Study 3: Digital Metering Data

This case study deals with data collected from customer water meters. The latest developments in technology enable water utilities to read the meters of their customer every hour or more frequently. This data provides valuable insights into how consumers use water, which allows water utilities to manage water resources better.

The chapters for this case study show how to analyse dates and times in R and use this to analyse water consumption. These insights are used to develop bespoke R functions that can be used to solve similar problems. The final chapter for this case study discusses some methods to detect anomalies in data, either by identifying bad data quality or unusual events that require a physical investigation.

1.2.4   Case Study 4: Predicting Concrete Strength

The last case study uses data from a machine learning repository that describes concrete mixtures. Concrete is a versatile material that is mixed from various components and commonly used in water management assets, such as filtration plants and tanks. The amount of water, cement, sand, gravel, additives, and the curing time determine the compressive strength.

This data is used to explain the principles of machine learning, by predicting concrete strength using multiple linear regression and a decision tree.

1.2.5   Closing Chapter

The final chapter in this book consolidates the lessons from this book and provides some advice on how readers can best continue to learn R and continue their quest to improve their data science chops.

1.3   What Is Data Science?

The idea that data helps us understand the world is thus almost as old as humanity itself. It has gradually evolved into what we now call data science. Using data in organisations is also called business analytics or evidence-based management. There are also specific approaches, such as Six Sigma, that use statistical analysis to improve business processes. Although data science is merely a new term for something that has existed for decades, some recent developments have created a watershed between the old and new ways of analysing a business. The difference between traditional business analysis and the new world of data science is threefold (Prevos, 2019):

1.   Businesses have much more data available than ever before

2.   Increased computing power

3.   Open-source innovation

This revolution is not only about powerful machine learning algorithms, but about a more scientific way of solving problems. Most analytical issues in supplying water or sewerage services do not require machine learning to solve. However, the definition of data science is not restricted to machine learning, big data, and artificial intelligence. These developments are essential aspects of data science but do not define the field.

One factor that spearheaded data science into popularity is the available toolkit, which has grown substantially in the past ten years. Open-source computing languages such as R and Python can implement complex algorithms previously in the domain of specialised software and supercomputers. As a result, Open-source software has accelerated innovation in how we analyse data and has placed complex machine learning within reach of anyone willing to make an effort to learn the skills.

Data science is a systematic and strategic approach to using data, mathematics, and computers to solve practical problems. A data scientist differs from a scientist who understands the world in generalised terms. The challenges facing data scientists are functional rather than scientific. A data scientist in an organisation is less interested in a generalised solution to a problem but focuses on improving how the organisation achieves its goals. In this sense, a data scientist is not strictly a scientist.

These three competencies are not the only qualities of a good data scientist. Other required skills are business competencies, such as writing a compelling business case, presentation skills, and understanding strategic and tactical objectives (Anderson, 2015). However, these skills apply to any professional working in a modern organisation and are certainly not unique to data science.

1.3.1 Data Science Unicorns

The best way to unpack the art and craft of data science is Drew Conway's often-cited Venn diagram (Figure 1.1). Conway defines three competencies that a data scientist or a data science team as a collective need to possess. The diagram positions data science as an interdisciplinary activity with three competencies: domain knowledge, mathematics, and computer science. A data scientist understands problems in mathematical terms and writes computer code to solve them.

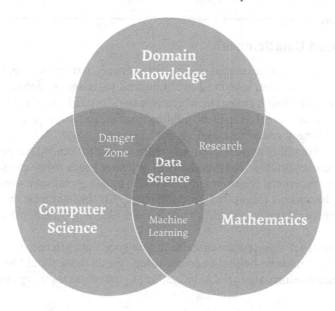

FIGURE 1.1 Conway Venn diagram.

The most vital skill within a data science function is *domain knowledge*. While the results of advanced applied mathematics such as machine learning are impressive, without understanding the reality these models describe, they are devoid of meaning and can cause more harm than good. Anyone analysing a problem must understand the context and potential solutions. The subject of data science is not the data itself but the reality this data describes. Data science is about things and people in the real world, not about numbers and algorithms.

The analyst uses *mathematics* to convert data into actionable insights. Mathematics consists of pure mathematics as a science and applied mathematics that helps us solve problems. The scope of applied mathematics is broad, and data science is opportunistic in choosing the most suitable method. For example, regression models, k-means clustering, and decision trees are some of the favourite tools of a data scientist. Combining subject-matter expertise with mathematical skills is the domain of traditional research.

To create value from electronic data, data engineers extract it from databases, combine it with other sources and clean the data before analysts can make sense of it. This requirement implies that a data scientist needs to have *computing skills*. Conway uses the term hacking skills, which many people interpret as unfavourable. Conway is, however, not referring to a hacker in the sense of somebody who nefariously uses computers, but in the original meaning of the word of a developer with creative computing skills. The core competency of a hacker, developer, coder, or whatever other preferable terms, is algorithmic thinking and understanding the logic of data structures. These competencies are vital in extracting and cleaning data to prepare it for the next step of the data science process.

Some critics of this model point out that people with all three skills are unicorns, mythical employees that can't exist in the real world. Most data scientists start from either mathematics or computer science, after which it is hard to become a domain expert. This book starts from the assumption that we can breed unicorns by teaching domain experts to write code and, where required, enhance their mathematical skills. Teaching water professionals to understand data science principles and write code helps an organisation embrace the benefits of the data revolution.

1.4 What Is Good Data Science?

Although data science is a quintessential twenty-first-century activity, we can find inspiration in a Roman architect and engineer, who lived two thousand years ago, to define good data science. Vitruvius wrote his book *About Architecture*, which inspired Leonardo da Vinci to draw his famous Vitruvian man. Vitruvius wrote that an ideal building must exhibit three qualities: *Utilitas*, *Firmitas*, and *Venustas*, or usefulness, soundness, and aesthetics (Prevos, 2019).

Buildings must have utility so people can use them for their intended purpose. A house needs to be functional and comfortable, and everybody in a theatre needs to see the stage. Each type of building has specific functional requirements. Secondly, buildings must be sound in that they are firm enough to withstand the forces that act upon them. Last but not least, buildings need to be aesthetic. In the words of Vitruvius, buildings need to look like Venus, the Roman goddess of beauty and seduction.

The Vitruvian rules for architecture can also define good data science. Excellent data science needs to have *utility*; it must be helpful to create value. The analysis should be *sound* so it can be trusted. Data science products also need to be *aesthetic* to maximise their organisational value.

1.4.1 Useful Data Science

Whether something is valuable is a subjective measure. What is helpful to one might be detrimental or useless to somebody else. For data science as a business activity, usefulness is the extent to which a data product contributes to the strategic or operational objectives. If a data science project cannot meet this criterion, then it is useless.

Meeting such goals in almost all cases means that the data we collect positively contributes to our ability to improve the world. The abstract activity of valuable data analysis directly or indirectly positively influences reality. When data science is separated from the world it seeks to understand or improve, it loses its power to be valuable.

After digesting a research report or viewing a visualisation, managers ask themselves: "What do I do differently today?" Therefore, data science's usefulness depends on the results' ability to positively empower managers and operational staff to influence reality. In other words, the conclusions of data science either comfort management that objectives have been met or provide actionable insights to resolve existing problems or prevent future ones.

Being data-driven is not the same as creating countless screens filled with backwards-looking dashboards (Anderson, 2015). Dredging through data to find something of value might be an exciting way to waste time. However, there is also a significant risk of finding fool's gold instead of valuable nuggets of wisdom. Therefore, the first step that anyone working with data should undertake before starting a project is to define the business problem that needs solving.

Data science is useful when the data product positively impacts reality, when new knowledge is created to solve future problems, or when we learn that the data can't answer the question that is asked (Caffo et al., 2018).

The well-known DIKW Pyramid (Data, Information, Knowledge, and Wisdom) explains how data produces a helpful analysis. Various model versions have been proposed, with slightly different terminology and interpretations (Frické, 2009). The modified version in Figure 1.2 better explains how data science can be helpful. Firstly, wisdom no longer forms part of the model because this concept is too nebulous to be beneficial. Instead, anyone seeking wisdom should study philosophy or practice religion, as data science cannot provide this need. Secondly, the bottom of the pyramid is grounded in reality. The standard DIKW model ignores the reality from which the data is collected, which leads to useless abstractions. The third modification to the traditional model is a feedback loop from knowledge to the real world. Anchoring the model for useful data science to reality emphasises the importance of domain knowledge when analysing data. In addition, subject-matter expertise helps to contextualise abstract data, resulting in better outcomes.

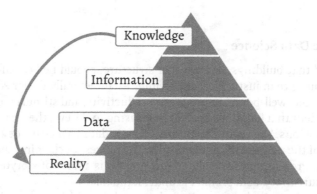

FIGURE 1.2 Alternative DIKW Pyramid.

The link between reality and data is not easy to maintain. Data does not always accurately describe the reality it measures or be complete as instruments need constant maintenance and calibration. Chapter 7 shows how to use R code to clean data collected from customer surveys. Anomaly detection is a technique that can detect issues with data quality, which is the topic of Chapter 12.

Moving from data to information is the core purpose of data analysis. Descriptive statistics (Chapter 4) summarises reality with a few numbers. But we only get genuine insights when transforming data into creating new insights, which is the topic of most chapters in this book.

Creating knowledge from information requires the analyst to present the findings in a way that helps the user of the data product to understand the reality from which the data originates better. A standard method to achieve this goal is visualising the data, which is the topic of Chapter 5.

The final step in the value chain requires the user of data science results to interpret the information to draw the correct conclusion about their future course of action. Good data science aims to ensure that the results of analysis lead to practical outcomes.

1.4.2 Sound Data Science

Just like a building should not collapse under its weight, so need the results of data science withstand the weight of critical review. Data science must be sound in that it is undertaken using best practice.

The Conway Venn diagram in Figure 1.1 shows that the overlap between computing and domain knowledge is the danger zone. We all have access to potent data tools, but wielding those without the appropriate mathematical nous will lead to analysis that is, at best useless and, at worse, wrong. The data and the analysis should be reliable and valid. Data science is reliable when it provides an accurate representation of reality. Validity relates to the extent to which measurements and the analysis describe the underlying reality. Chapter 8 discusses these topics in more detail concerning customer surveys.

The second important aspect of sound data science is its reproducibility and replicability, which is the ability of other people to reconstruct the analyst's workflow from raw data collection to reporting. This requirement is a distinguishing factor between traditional business analysis and data science. Reproducible data science uses the same code and data to verify the assumptions and results. Replicability is about repeating the same analysis with different data. Reproducing an analysis is the cornerstone of quality assurance. Replicating an analysis is important in business automation to increase efficiency. Chapter 6 shows how to generate a PowerPoint presentation using R code from raw data.

Water data science is typically about data collected from electronic devices, where our main concern is ensuring the measurements are calibrated. However, as water utilities collect more data about their customers, there is also a need to adhere to ethical principles, further explained in Section 1.4.4.

1.4.3   Aesthetic Data Science

Vitruvius insisted that buildings, or any other structure, should be beautiful. The aesthetics of a building causes more than just a pleasant feeling. Architecturally designed places stimulate our thinking, increase our well-being, improve our productivity, and stimulate creativity.

While it is evident that buildings need to be pleasing to the eye, the aesthetics of data products might not be so obvious. The requirement for aesthetic data science is not a call for beautification and obfuscation of the ugly details of the results. The process of cleaning and analysing data is inherently complex. Presenting the results of this process is a form of storytelling that reduces this complexity to ensure that a data product is understandable.

Aesthetic data science is about creating a data product, which can be a visualisation or a report designed so that the user draws correct conclusions. A messy graph or an incomprehensible report limits the value extracted from the information. Visualisations that are not aesthetic are hard to interpret, which leads to the wrong conclusions or even deceives the viewer (Machlis, 2019). Chapter 5 explains some principles of visualising data to tell stories and how to implement this with R.

1.4.4   Data Science Ethics

Collecting sensitive customer information has ethical implications because it reveals information that might impact the customer when accidentally released or used in analysis. Most water data is from material sources and thus has no ethical implications. Still, when collecting data from customers, we need to consider ethics beyond compliance with privacy legislation.

Traditional social research has well-developed ethical processes for gathering, storing, and interpreting information that describes people's lives. Social scientists have developed these principles in response to negative experiences with unethical research. Three crucial ethical principles that apply to data science are (Bryman & Bell, 2011):

- Informed consent

- Avoiding harm in collecting data

- Doing justice to participants in analysing data (algorithmic fairness)

When we ask people to complete surveys, subjects can choose which information they provide or abstain from providing information about themselves. The principle of informed consent ensures that participants are neither deceived nor coerced into surrendering information about their personal lives.

Harm in collecting data includes physical or psychological consequences for the participant, such as their development or self-esteem, inducing stress and introducing risk to career prospects or future employment. Most countries have well-established privacy legislation that defines how organisations should deal with personal information.

Doing justice to participants implies that any analysis and interpretation of customer information reflects their interests. Algorithmic fairness should prevent potential harm to people through analysis. Machine learning algorithms use information from the past to predict possible futures. Because algorithms are anchored in the past, any machine learning algorithm will amplify any bias within the training data. The second aspect of algorithmic fairness relates to an imbalance in power between the subject of data science and the organisation that uses the algorithm. For example, algorithms can maximise the amount customers pay for a service. Although perfectly legal, this is only possible due to an imbalance in power between the seller and the buyer (Davis & Patterson, 2012).

Data obtained from digital customer water meters can reveal how many people live in a house, their nightly toilet habits, when they are on holiday, and so on. Unfortunately, we could use this data to negatively impact customers by maximising prices or for mass surveillance.

2

Basics of the R Language

Now let's roll up our sleeves and write some code. This chapter introduces the principles of the R language and RStudio to analyse and visualise data. This chapter finishes with a case study to convert level measurements in an open irrigation channel into flows. The learning objectives for this chapter are:

- Install R and RStudio and identify the different parts of the RStudio screen

- Understand the principles of writing code to analyse data

- Apply R code to solve a simple water problem

2.1 Download and Install R and RStudio

The best way to use R is through an *Integrated Development Environment* (IDE). This type of software helps you to write and manage code. An IDE typically consists of a source code editor, automation tools, and functionality to simplify crafting and running code. Of course, you can use R without the IDE, but it will be a bit less user-friendly.

Several IDEs are available to help you write R code. RStudio by Posit is the most popular option. This software is also an open-source project, with free and paid versions for companies that want to use advanced features and support services. RStudio can also work with other languages such as SQL and Python. Follow these steps to install the required software:

1. Go to the R Project website: `cran.r-project.org`
2. Download the *base* version for your operating system and install the software
3. Go to the download page on the RStudio website: `posit.co`
4. Download the installer for the free desktop version and install the software

If you use Linux, you can use your package manager to download and install R and RStudio. Windows users should also install *rtools4*, available from the CRAN website, which provides advanced functionalities for managing R and its packages. If you are not using your own computer, check with your administrator to obtain relevant access to the system. Advanced security measures such as whitelisting will make it hard to use the desktop version on your computer.

Alternatively, you can sign up for a free and fully featured account to access RStudio's cloud version (`posit.cloud`). This service gives you full access to R and RStudio in your browser without installing any software. The cloud version has the same functionality as the desktop version. The free version provides enough hours of computing time to work through this book. You'll have to pay for a subscription or install the desktop version if you need more time. When you open the desktop version of RStudio, you are ready to go, but with the cloud version, you will need to start a new project before you see the screen in Figure 2.1.

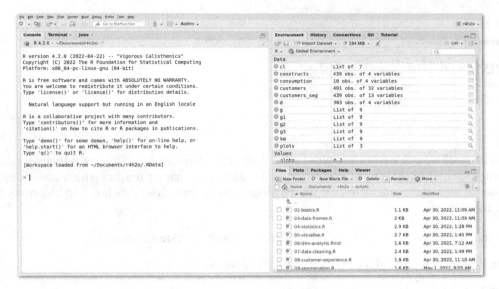

FIGURE 2.1 RStudio IDE interface (Source: posit.cloud).

2.2 Basics of the R language

2.2.1 Using the Console

We start with the console or the REPL (Read-Eval-Print Loop). The console is like a desktop calculator which prints results on a paper roll. The result will appear in the console below your code as you
issue instructions. The console is the left screen within RStudio that displays the R version you are
working with and a greater-than sign at the bottom of the frame. Move your cursor to the console
(or type Control-2) and enter the code examples below. For ebook readers, don't copy and paste the
examples because typing the code develops your muscle memory for the R syntax and introduces
you to some of the features of the IDE. The greater-than sign at the start of each line in the console
is the prompt that indicates the currently active line. This symbol tells you where the cursor is. The
examples in this text do not show the prompt. Let's start with a simple example to calculate the area
of a pipe. Type this code into the console and hit ENTER after each line:

```
diameter <- 150
diameter
```

```
[1] 150
```

```
(pi / 4) * (diameter / 1000)^2
```

```
[1] 0.01767146
```

You have now written your first R code! You will notice a few things as you enter the code:

- When you hit enter, the result of expressions without the assignment symbol (<-) is shown in the
  console.

- The number between square brackets [1] indicates that this is the first and, in this case, the only
  result.

- RStudio includes the closing bracket or quotation mark when typing parentheses or quotation marks.

- The variables you declare (`diameter` and `pipe-area`) and their values are shown in the *Environment* window.

- The variable `pi` is a built-in constant, accurate to 16 decimals.

If you want to rerun a line of code or you like to edit it, use the up and down arrow keys to browse the console history. Your history is also available in the *History* tab in the top-right window. When you save the project, RStudio saves the history to disk. You can clear the history of your session with the broom icon.

2.2.2   The Assignment Operator

R uses the `<-` operator, a left-handed arrow with the minus and lesser-than symbol, to assign values to a variable. For example, a `<- 6` assigns the number 6 to the variable a. This notation means the value 6 or R is placed into the variable a. Note that the assignment operator works both ways, so a `<- 6` is equivalent to `6 -> a`. When we look into functions in Chapter 12, we also use the «- operator to assign variables. You can use the = symbol, which is common in other languages, but this method can lead to confusion when writing more advanced code, albeit rarely. Operator conflicts are rare and you can safely use the = symbol in almost all cases, but correct R code uses the official assignment operator, even if it is just a matter of style.

2.2.3   Variables Names

Variables are the basic building blocks of computational analysis. A variable can store numbers, text, matrices, or any other information that needs to be analysed or presented. In a spreadsheet, a variable is a cell or a group of cells, while in data science code, variables have names.

You can give variables almost any name you like, as long as they only contain letters, numbers, dots, and/or underscores and start with a letter. You cannot use arithmetic symbols in variable names. For example, when you evaluate `flow-daily`, R will try to subtract the variable `daily` from `flow` instead of creating a new variable. Try to use a meaningful name that describes its content. For example, don't call a flow measurement `f`, but `flow_daily` or something similar. The first word usually describes what the variable means (in this case, `flow`), and the remainder add qualifiers (such as `daily` or `maximum`). Data scientists most commonly use one of five forms to write long variable names (Bååth, 2012):

- `flowdaily`: All lowercase

- `flow.daily`: Period-separated

- `flow_daily`: Snake case

- `flowDaily`: Camel case

- `FlowDaily`: Upper camel case

There are no strict rules on how to name variables. It depends a lot on your native language and personal preference. Whatever method you use, try to be consistent so that your code becomes easy to understand when you share it or read it months later.

Don't worry about having to type too much. Soon after entering three characters, R shows every possible variable name or function that starts with the entered letters in a little popup menu.

Continue typing or use the arrow keys to pick the option you need and hit the TAB key to complete. Autocompletion saves time and is a great incentive not to use cryptic variable names.

All variables you declare are listed in the *Environment* tab in the top-right window. You can remove variables with the `rm()` function. For example, using `rm(flow_daily)` will remove that variable. To clear the whole environment when things get messy or you run out of memory, use `rm(list = ls())`, which removes the list of all variables. Alternatively, you can click the broom icon in the Environment tab.

2.2.4 Doing Arithmetic

In its most basic form, the R console is a calculator that uses familiar arithmetic operators (+, -, *, /). R uses the caret symbol ∧ for exponentiation. The modulo (%%) function returns the remainder of a division: `17 %% 5` results in 2. Integer division (%/%) returns the truncated result of a division: `17 %/% 5` results in 3. Furthermore, R applies the BODMAS rule to find the correct answer to the annoying arithmetic memes distributed on social media.

```
3 - 3 * 6 + 2

[1] -13

7 + 7 / 7 + 7 * 7 - 7

[1] 50
```

2.2.5 Vector Variables

The variables we defined above are scalars with only a single value. Data is, however, rarely a single point but stored in a vector or matrix. Vector arithmetic is the most essential principle in R. A vector is a sequence of variables of the same type, which can be defined with the collection function: `c()`. The colon is a shortcut to combining observations in a vector. For example, the expressions `1:3` and `c(1, 2, 3)` are identical. This functionality is often useful when creating indices.

```
complaints <- c(12, 7, 23, 45, 9, 33, 12)
day <- 1:100
```

2.2.6 Arithmetic Functions

A series of simple functions is available to undertake basic mathematical operations. R functions can also perform complex tasks such as visualising and analysing data. A function call consists of a word and parenthesis, such as `sqrt()`, to determine the square root of a number or variable: `sqrt(25)`. Most functions have one or more parameters between parentheses, but some take none. Functions apply to both scalar variables and vectors. This method makes it easy to use a mathematical operation on a large set of numbers with one line of code. You can, for example, call `sqrt(c(1, 4, 9, 16, 25))` to obtain a new vector with the square roots of these five numbers. Vector arithmetic is an essential principle of the R language. Table 2.1 shows some of the essential mathematical functions available in R. Most of these functions only need a vector of numbers to provide a result. Functions, such as `log()`, can also have parameters to control the process. The logarithm function results in the natural logarithm by default, which can be changed with the `base` parameter. For logarithms with base ten, use `log10()`. Let's apply these functions to a vector of flow measurements and inspect the results.

TABLE 2.1 Arithmetic functions.

Formula	Description	Function
$\sum x$	Sum of all elements	`sum(x)`
$\prod x$	Product of all elements	`prod(x)`
$x!$	Factorial	`factorial(x)`
e^x	Exponential	`exp(x)`
$\log_y x$	Logarithm	`log(x, base = y)`
$\log_{10} x$	Ten-base logarithm	`log10(x)`
$\sqrt{x}$	Square root	`sqrt(x)`
$|x|$	Absolute value	`abs(x)`

```
non_revenue_water <- c(13, -9, 45, 0)

sum(non_revenue_water)

[1] 49

prod(non_revenue_water)

[1] 0

factorial(non_revenue_water)

[1] 6.227021e+09        NaN 1.196222e+56 1.000000e+00
Warning message:
In gamma(x + 1) : NaNs produced

exp(non_revenue_water)

[1] 4.424134e+05 1.234098e-04 3.493427e+19 1.000000e+00

log(non_revenue_water)

[1] 2.564949      NaN 3.806662      -Inf
Warning message:
In log(non_revenue_water) : NaNs produced

sqrt(non_revenue_water)

[1] 3.605551      NaN 6.708204 0.000000
Warning message:
In sqrt(non_revenue_water) : NaNs produced

abs(non_revenue_water)

[1] 13  9 45  0
```

Some calculations resulted in either NaN or Inf. The Inf or -Inf indicator appears when the calculation results approach positive or negative infinity. The NaN indicator means Not a Number, which occurs when you apply the factorial, logarithm, or square root function to a negative number. As an aside, R can work with complex numbers, but you need to indicate this with the `as.complex()` function: `sqrt(as.complex(-25))`. Alternatively, you can nest the `abs()` function to turn the values positive (`sqrt(abs(-25))`).

You have written a bit of code now, and if your console is cluttered, you can type Control-l to clear the screen or click on the broom button in the console frame.

2.2.7   Basic Visualisations

R has extensive capabilities for visualising data and the results of your analyses. The code snippet below generates a simple graph shown in Figure 2.2. The first lines of the code define the variables. The plot function visualises these values with some additional parameters. Next, the `abline()` functions draw straight lines on the current plot and `points()` adds the marker. Plotting functions have a lot of possible parameters, such as colour (`col`), line type (`lty`) or character type (`pch`), and size (`cex`).

```r
diameters <- 50:351
pipe_areas <- (pi / 4) * (diameters / 1000)^2

plot(diameters, pipe_areas,
     type = "l", col = "blue",
     main = "Pipe Section Area",
     xlab = "Diameter", ylab = "Pipe Area")
abline(v = 150, col = "grey", lty = 2)
abline(h = (pi / 4) * (150 / 1000)^2, col = "grey", lty = 2)
points(150, (pi / 4) * (150 / 1000)^2, col = "red", pch = 12, cex = 2)
```

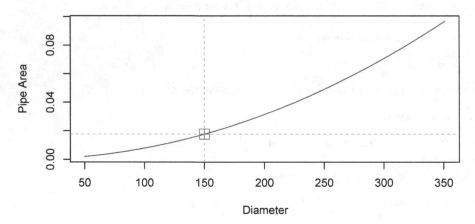

FIGURE 2.2 Basic plotting functionality.

The graph appears in the *Plots* tab. This tab contains a history of all plots through the arrow buttons. You can also export the graph to a file or copy it to a clipboard. This graph is only a simple example of visualising data. Chapter 5 discusses this topic in more detail.

The best way to learn R code is to play with code somebody else wrote. Experiment with different options in this code and review the result. Using the TAB key after a comma inside the function, a help window with available options appears. Use the arrow and enter key to navigate the completion menu.

2.3 Developing R Code

The console provides a running record of the expressions that R evaluates. While this method is helpful for quick calculations, using the console makes it hard to reconstruct what steps you have taken to get to your result. Referring back to the principles of good data science, the console is not reproducible.

You write your code in an R script to create reproducible code. A script is a text file with lines of code saved to disk. You can open this file in any text or word processor, so your work can be read by anyone with a computer.

Create a new R script by going to *File > New File > R Script* or hitting `Control-Shift n`. You can also open an existing file from the same menu or with `Control o` shortcut. The code accompanying this book contains scripts for each chapter and case studies for you to explore.

R does not evaluate the code when you hit enter within a script. Instead, the editor adds a new line, just like any word processor or text editor. To execute a line of code in the editor, you need to type `Control-Enter` or click the *Run* button on top of the frame. You can also select a section of code or part of a line and only run that part. The result will be shown in the console or the graphics frame. When you hit the *Source* button or press `Control-Shift Enter`, RStudio evaluates all code in the script.

2.3.1 RStudio Projects

After a while, your drive will be littered with R scripts and data files. RStudio helps you organise this chaos using projects. An RStudio project is a set of files that relate to each other. Each project has a working directory, workspace, history, and source documents. Every time you open a project, it is in the same state where you left it when you last closed the program.

In the cloud version of RStudio, your workspace lists all available projects. Select *New Project > New Project from Git Repository* to access the course data. Git is a commonly used program by developers to manage file collections. You can access the code and data related to this book by entering: `htps://github.com/pprevos/r4h2o/`. Wait for the files to be copied.

Desktop version users can download the book files by going to the GitHub website, clicking the code button, and then on *Download ZIP*. Download and extract the files to your computer, and you are ready to go. In the desktop version, select *File > Open Project* and select the `r4h2o.Rproj` file. You should see a list of files, as shown in the bottom-left frame in Figure 2.1, in both the desktop and online versions.

2.3.2 Writing Elegant Code

For code to be reproducible, it needs to be written elegantly. To help with this task, software developers have developed a set of best practice roles to make a computer script easier to follow. Computer science guru Donald Knuth said in this respect that computer code is like poetry (Knuth, 1974).

Looking at a blob of plain text can be challenging because text editors have no special fonts, bold, italic, or lines to draw the reader's attention. Instead, RStudio and other IDEs use syntax highlighting in your script to make the code easier to read. Functions, comments, strings, and numbers have different colours to stand out from the rest of the script. The colours have no function other than help you navigate the script. Text editors use themes that define these colours. You can change your theme by going to *Tools > Global Options > Appearance* and selecting your favourite. Most developers prefer themes with dark backgrounds as they are easier on the eye when staring at a screen for a long time.

R functions can become long as you add more parameters. When you use very long lines, code can be hard to read. The RStudio script panel shows a vertical line beyond which, ideally, you should not write any code. You can break an instruction over multiple lines to enhance its readability. Best practice is to split a line at a comma, as shown below. The text editor will indent the code, making it easy to see which lines belong together, like in the code from the previous section.

```
# Example of nicely formatted code
plot(diameter, pipe_areas,
     type = "l", col = "blue",
     main = "Pipe Area")
```

Adding space around every operator and after each comma is good practice. Trying to read large chunks of code without much whitespace can be daunting. Don't be afraid of empty space in your code. Don't suffer from *horror vacui*, which applies to visual artists who seek to fill all empty space in their works of art. Adding these spaces will make your code less busy and easier to read.

Looking at a screen full of code can be daunting at the best times. To guide the reader through the code, developers use comments. A comment is a statement that is not evaluated when running the code. In the R language, comments are indicated with one or more pound signs #, also known as a number sign, hash, or hashtag, at the start of a line or after a function call.

There are, of course, no absolute rules when it comes to how you write your code. If you insist, you can write a whole program in one single line, but it will be sheer impossible to comprehend the code when you get back to it. There is no overarching style guide for the R language, but some coding standards have been published to guide developers and achieve internal consistency (Bååth, 2012). All style guides are, however, fundamentally opinionated as there is no absolute truth in coding style. The most crucial objective for any coders is consistency, whitespace, and adding comments.

2.3.3   Finding Help

While mastering the syntax of R might seem daunting, the RStudio development environment helps you with writing code. The R language has a built-in help function for every function. For example, type `help(sum)` or `?sum` to learn everything about this function. Unfortunately, these help files can be cryptic to beginning users, but they all follow the same structure. The first section of the help file describes the function in words. The second section shows how to use the function. Finally, the third section lists the arguments of the function. The following sections in the help function provide background information and links to other similar functions. Most help entries also show examples that help you to reverse-engineer the functionality. Many users jump to the examples to see how the function should be used.

R will print an error message or a warning when you evaluate syntactically incorrect or incomplete code. Instead of being called "error messages", they should be called "help messages". It is often helpful to paste the entire error message into a search engine because somebody else will have most likely asked a question about it previously.

2.3.4   Debugging Your Code

Writing computer code is at the same time rewarding and frustrating. Achieving the desired outcome provides an incredible feeling of accomplishment. But this sense of achievement comes with challenges. The biggest problem with writing code is that the computer executes your instructions exactly as they are written. Therefore, any ambiguity in the code will lead to unexpected outcomes. These are the famous bugs that can be frustrating and time-consuming to eradicate. The very first computer bug ever discovered was an actual bug. Early computers were the size of a small room, and in 1947 some engineers found that moths would short-circuit the wiring.

When the results of your analysis don't make sense, or the code doesn't work at all, always blame yourself and not the computer. Even a tiny mistake, such as a misplaced bracket or wrong case (for example, `Plot()` instead of `plot()`), causes the program to fail. The best way to prevent bugs is to be systematic and write elegant, easy-to-understand code.

Syntactic errors are common and easily caught. RStudio will help you find simple semantic errors, indicated with clickable markers in the margins of the code. However, even though your code might be syntactically and semantically correct, you might not achieve the expected outcome. Either the program crashes or runs forever in an infinite loop. If R produces an error message:

- Check for typos. A parenthesis in the wrong place or mixed uppercase, and lowercase can cause a lot of trouble.

- Read the help file for the relevant function.

- Search the error message, precisely as written, with your favourite search engine.

More problematic is when your code provides the wrong answer. To prevent these errors, it is wise to test your code with known data where you can anticipate the outcomes. Remember to always undertake a common-sense review of your results.

2.3.5 Reverse-Engineering

Not all code examples in this book are explained in detail. To understand how the code functions, you need to reverse-engineer it. Modifying existing code to figure out how it works is a productive method to playfully learn a programming language. The easiest method to reverse-engineer code is to execute each line or part of a line separately and inspect the intermediate results. Another technique is changing function parameters and analysing differences in the output. When you search the internet, you find many examples of code shared by others in their blogs or data science websites. Copying code from the web is an efficient way to learn to code. Analyse them and use the help files to learn new techniques and replicate them in future problems.

2.4 Case Study

Now it is time to solve a water problem. Measuring the flow through a pipe or an irrigation channel is one of the most fundamental metrics in water management. This case study calculates the flow through an open channel with basic measurements.

Some water utilities use open channels to transport water between supply systems or rural customers. These channels use gravity to transport water downhill, just like the Roman aqueducts in Europe or Balinese Subaks. The flow through these channels is given by their geometry and the slope.

Determining the flow in an open channel is usually achieved by measuring the water depth through a section with a known shape. Forcing the water through a given shape over a sharp weir creates a boundary condition that allows us to calculate the flow. A mathematical relationship determines the volume of water that passes through the channel.

This case study uses a rectangular weir, as shown in Figure 2.3. A simplified version of the Kindsvater-Carter rectangular weir formula (Equation 2.1) describes the flow in SI units (ISO 1438, 2017).

$$Q = \frac{2}{3}C_d\sqrt{2g}\,bh^{(3/2)} \qquad (2.1)$$

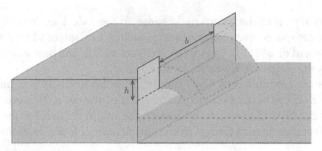

FIGURE 2.3 Rectangular weir.

- Q: Flow rate (m³/s)
- C_d: Discharge coefficient
- g: Gravitational acceleration (9.81 m/s²)
- b: Measured width of the notch [m]
- h: Upstream head above crest level [m]

The value for C_d is an estimate because it depends on the dimensions of the channel and the flow characteristics. With this information, we can answer some questions.

1. What is the flow in the channel in m³/s when the height $h = 100$mm?
2. What is the average flow for these heights: 150mm, 136mm, and 75mm, in l/s?
3. Plot the flow in m³ per second for all heights (h) between 50mm and 500mm.

For this case study, lets use the following assumptions:

- $C_d = 0.62$
- $g = 9.81$ m/s²
- $b = 0.5$ m

Before doing any calculations, you create a new R script and define all variables you know as constants:

```
cd <- 0.62
g <- 9.81
b <- 0.5
```

To evaluate the Kindsvater-Carter formula, use the `sqrt()` function to calculate the square root. The key to getting the formula right is to use parentheses where appropriate. The dimensions in the formula are in metres, while the measurements are in mm. You thus need to use `h1 / 1000` in your formula. The input is a single scalar variable, and all the constants are scalar; therefore, the output is also a single number.

```
h1 <- 100 / 1000
q1 <- (2 / 3) * cd * sqrt(2 * g) * b * h1^(3 / 2)
q1

[1] 0.02894809
```

To answer the second question, we first create a vector of height measurements with the `c()` function, so you use the formula only once. Finally, the `mean()` function calculates the total volume.

```
h2 <- c(150, 136, 75) / 1000
q2 <- (2 / 3) * cd * sqrt(2 * g) * b * h2^(3 / 2)
mean(q2)
```

```
[1] 0.03929853
```

The last question follows the same approach as the second one. First, create a vector, calculate the flow and plot the results. For good measure, the plot adds some lines to indicate the solutions to question 2 (Figure 2.4).

```
h3 <- (50:500) / 1000
q3 <- (2 / 3) * cd * sqrt(2 * g) * b * h3^(3 / 2)

plot(h3, q3, type = "l",
     xlab = "Height [m]", ylab = "Flow [m3/s]",
     main = "Open Channel Flow, Cd = 0.62")
abline(v = h2, col = "grey")
abline(h = q2, col = "grey")
points(h2, q2, pch = 19, col = "blue")
```

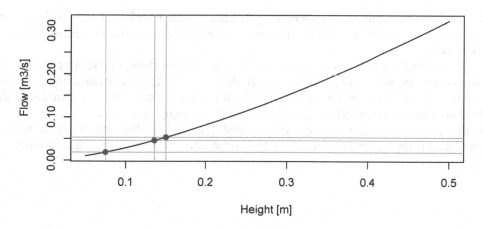

FIGURE 2.4 Kindsvater-Carter formula visualised.

2.4.1 Generating Sequences

R also has a special function to generate sequences. Evaluating `seq(from = .05, to = .3, by = .001)` gives the same result as `(50:300) / 1000`. You can also specify a length, for example `seq(from = 0.05, to = 0.3, length.out = 100)` results in a vector of 100 equally-spaced numbers.

```
h3 <- seq(from = .05, to = .3, by = .001)
```

```
h3 <- seq(from = 0.05, to = 0.3, length.out = 100)
```

2.4.2 Repeat Code with Loops

Another method to automate this process is a loop. The first line defines an empty vector with the name q. The for-loop runs from 50 to 500 and stores the results in the vector. The variable i provides the index of the vector. Note that in R, the first element of a vector has position 1, while in almost all other program languages, vectors start at index zero.

```
q3 <- vector()
i <- 1

for (h3 in 50:500) {
  q3[i] <- (2 / 3) * cd * sqrt(2 * g) * b * (h3 / 1000)^(3 / 2)
  i <- i + 1
}
```

While most other programming languages frequently use loops, R uses vector arithmetic. Therefore, loops should be avoided when possible as they are slow when running a large number of iterations. However, in some situations, loops are unavoidable, so you do need to know how to use them.

2.5 Further Study

Learning a new programming language means you need to remember lots of new vocabulary and syntax. As a result, most data science coders regularly use reference materials to remember how to undertake a task.

The good people from RStudio have developed a series of cheat sheets that summarise valuable functions and syntax, which are available through their website. The *Base R Cheatsheet* provides a comprehensive list of the basic functionality in the R language. The course associated with this book also has a cheat sheet, which is available on github.com/pprevos/r4h2o.

Now that you are warmed up, we can start analysing some data. The next chapter introduces how to read data from files and explore its content using a case study with water quality data.

3

Loading and Exploring Data

We have covered the basics of the R language, so it is time to load and explore some data. The first action in any analysis project is to obtain and load data. This data is available in a database, the cloud, or a file. One of the essential skills is thus to load this data into memory and view its content. The subsequent four chapters use water quality data from an imaginary water supply network, which we analyse for compliance with regulations. This chapter has the following learning objectives:

- Download and install R packages

- Load and describe a CSV and Excel data

- Explore the content of rectangular data (data frames)

3.1   R Packages

When you open R, it has a lot of built-in functionality to undertake complex statistical analysis. The basic R language is versatile, but some specialised tasks require additional packages. Packages are like apps that extend the capabilities of your phone. A large community of users develop R code and make it freely available to others. As a result, thousands of specialised packages are available that help you undertake a broad range of complex tasks. You can, for example, use R as a Graphical Information System (GIS) and analyse spatial data or implement machine learning. Other packages help you access data from varied sources, such as SQL databases. R packages can also help you present the results of your analysis as a presentation, report, or interactive dashboard.

Most R packages are available on CRAN (`cran.r-project.org`), the *Comprehensive R Archive Network*. You can install packages in R with the `install.packages()` function. Within RStudio, you install packages in the *Tools* menu. When you install a package, R downloads a library of files and stores them for future use on your computer. The words package and library are sometimes used interchangeably. A package is a collection of R functions, data, and documentation. The location where the packages are stored is called the library. The CRAN repository contains many packages with specific functionality to analyse water, some of which are:

- *baytrends*: Long-term water quality trend analysis (Murphy et al., 2022)

- *biotic*: Calculation of Freshwater Biotic Indices (Briers, 2016)

- *EmiStatR*: Emissions and statistics in R for wastewater and pollutants in combined sewer systems (Torres-Matallana, 2021)

Exploring the thousands of packages that R offers can be a daunting experience. Most online searches lead you to the CRAN website, which hosts the most extensive collection of packages. Each package has a PDF reference manual that lists all the help files. These documents can, unfortunately, be a bit cryptic.

Many packages also include vignettes, which provide a more narrative introduction to the added functionality. For example, the *dplyr* package includes several vignettes. When you install a package, the vignettes are downloaded to your system. You can view it with `vignette("dplyr")`. Without a parameter, this function shows a list of all available vignettes on your system. The `package` parameter shows all vignettes for a package: `vignette(package = "dplyr")`.

In the desktop version, you only need to install a package once. However, in the cloud version of RStudio, you will need to install packages for each project. As you use your packages for a while, they might go out of date because developers have developed new versions. The `update.packages()` function checks which ones are outdated and asks whether you like the latest version. You can find a list of all packages installed on your computer in the *Packages* tab of your RStudio screen.

After installing a package, you need to initialise it with the `library()` function. You need to do this in every script so that R knows which packages you want to use in this instance.

3.1.1   Introducing the Tidyverse

One of the most popular collections of R packages for analysing data is the *Tidyverse*, developed by R guru Hadley Wickham and many others (Wickham, 2022c). Doing Tidy data science is a coding style with a strong following. Tidy data science relates to writing code and cleaning data in a way that promotes reproducibility and clarity of the code.

The Tidyverse collection of packages provides enhanced functionality to extract, transform, visualise, and analyse data. In addition, the features offered by these packages are more versatile and easier to use and understand than the base R code. Computer scientists call software like the Tidyverse syntactic sugar, which means that it simplifies the syntax and sweetens the process of analysing data. Most Tidyverse functions have equivalents in base R, but they are a bit easier to use, which is why this book features this package.

Installing the complete Tidyverse can take a little while. If you are working with the desktop version, ensure you have sufficient access rights to install and run the packages.

```
library(tidyverse)
```

When you initiate the Tidyverse library for the first time, R shows feedback in the console that eight packages are loaded. You can also activate each of these packages individually, for example `library(readr)` or `library(dplyr)`.

- *dplyr*: Manipulate and analyse data (Wickham et al., 2022b)

- *ggplot2*: Visualise data (Wickham et al., 2022a)

- *forcats*: Working with factor variables (Wickham, 2022a)

- *purrr*: Functional programming (Wickham & Henry, 2023)

- *readr*: Read and write read rectangular data (Wickham et al., 2022c)

- *stringr*: Manipulate text (Wickham, 2022b)

- *tibble*: Working with rectangular data (Müller & Wickham, 2022)

- *tidyr*: Data transformation (Wickham et al., 2023)

You can ignore the message after you activate the Tidyverse libraries. These warnings relate to functions that override existing functionality and are not a concern for now.

When you use a function from a package without first calling it with the `library()` function, R will not recognise it or use another one with the same name from the base package. Furthermore, you can use functions from packages without calling the library by adding the library name in front of the function, followed by two colons, e.g., `readr::read_csv()`. You only need to use this method to resolve conflicts or when you only use a package once.

The Tidyverse developers frequently update the software. You can see if updates are available and optionally install them by running `tidyverse_update()`. You can also upgrade packages in the *Tools > Check for Package Updates* menu in RStudio.

These are not the only packages that follow the Tidyverse principles. The official repositories contain more than twenty packages. In addition, many other developers also follow the principles of the Tidyverse, such as *Tidytext* (Silge & Robinson, 2016) for text mining and *anomalize* for anomaly detection (Dancho & Vaughan, 2023).

3.2   Case Study 1: Exploring Water Quality Data

The first case study looks at water quality data in four towns in the fictional city of Gormsey. Each suburb has a set of sample points at randomly selected customer taps. Gormsey's laboratory regularly samples these taps and tests the water for various parameters. Water utilities test for hundreds of different parameters at different frequencies at the water treatment plant and in the water network. Analysing this type of data is an everyday activity for water professionals. However, the data for this case study only contains four parameters. This file contains the results from samples taken at locations in the network for:

- Turbidity

- Escherichia coli (E. coli)

- Total chlorine

- Trihalomethanes (THM)

Turbidity is a measure of the cloudiness or opaqueness of a liquid. It is a measure of the aesthetics of drinking water and an indicator of possible further issues. The unit for turbidity is NTU or Nephelometric Turbidity Unit. The instrument incorporates a single light source and a photodetector to sense the scattered light.

E. coli bacteria are found in the natural environment, foods, and intestines of people and animals. Although most strains of E. coli are harmless, some can make you sick and can cause diarrhoea, while others cause urinary tract infections, respiratory illness, and other diseases. E. coli in freshwater is measured by counting the number of colonies incubated at 35°C for a day. This bacteria provides direct evidence of faecal contamination from warm-blooded animals and is an essential indicator in water quality management.

Chlorine is an effective control against the presence of bacteria in drinking water. It is usually dosed at the water treatment plant or in the network. Chlorine is broken down into other components as it travels through the network. The primary consideration for managing a water network is that chlorine levels are sufficient to disinfect the water. But the chlorine levels should not be too high. High chlorine levels dissuade people from drinking tap water, motivating them to use less healthy alternatives.

THMs are a group of chemical compounds predominantly formed as a by-product when chlorine is used to disinfect drinking water. Long-term exposure to a high level of these by-products can cause diseases such as cancer.

This case study uses simulated data. All dates are shifted into the distant future, and the names of locations are fictional. Most real data is reliably dull as water utilities manage water quality within acceptable limits. Therefore, some results have been exaggerated to create some interesting outcomes.

The data for this chapter are available in the data folder of your RStudio project. Section 2.3.1 describes how to obtain this data.

3.3   Loading Data

Raw data is the lifeblood of analysis which can come in many forms. Packages are available to read almost every imaginable data format. You can read and write to databases, scrape data from the web, and read a broad range of file formats, including spatial data, from disk or directly from the internet.

So far, we have seen two types of variables. Scalars contain a single value, and vectors contain two or more values. Adding one more dimension, we enter the world of matrices. The most common form of two-dimensional data in R is the data frame. In a data frame, all columns are variables, and each variable can be of a different kind, such as a number, text, or date. The rows hold the observations (Figure 3.1). This data is often available as a table in a database, JSON files (JavaScript Object Notation) on the internet, spreadsheets, or CSV files (Comma-Separated Values).

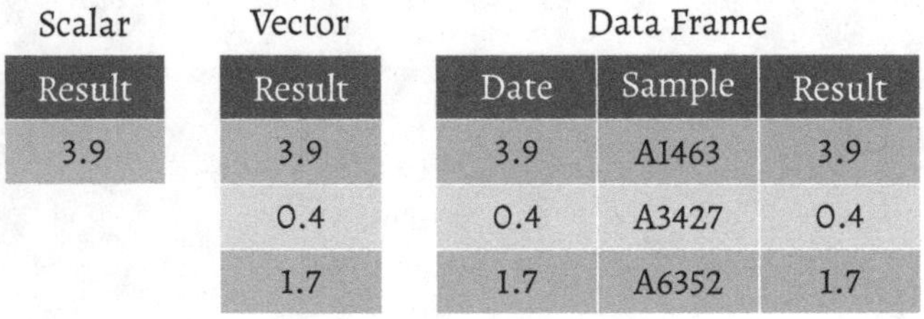

Scalar	Vector	Data Frame		
Result	Result	Date	Sample	Result
3.9	3.9	3.9	A1463	3.9
	0.4	0.4	A3427	0.4
	1.7	1.7	A6352	1.7

FIGURE 3.1 Scalars, vectors, and data frames

3.3.1   Reading CSV Files

This book uses CSV files to provide access to data, which R can read using the read.csv() function. The output of this function is a data frame. This function loads the content of the CSV file into memory. The file's content does not change unless you instruct R to write the data to disk with the write.csv() function. Best practice in data science is to not change the raw data to ensure reproducibility of the analysis.

```
labdata <- read.csv("data/water_quality.csv")
```

The text between quotation marks is the path to the file and its name. R uses the forward-slash (/), common in Unix systems, and not the Windows backslash (\) to form a path. Every R session has a working directory, and all paths are relative to that folder. RStudio saves the active directory for future sessions when you work on a project. You can see the current working directory with the getwd() function. Without a working directory, you would need to specify the complete path, such as C:/Users/peterp/Documents/r4h2o/data/water_quality.csv. You don't need to remember your filename exactly, or which folder it lives. Instead, when you start typing a file name, RStudio will use its auto-completion to help you find the file. The *Files* tab in the bottom-right of your screen shows all folders and files in your working directory. You can use this screen to open and manipulate files.

The labdata variable is shown in the environment tab after you evaluate the code. This tab shows the number of observations (rows) and variables (columns). You can click on the triangle to expand the entry, which shows the variables, their type, and the first few values. Clicking on the variable name opens a new frame that shows the data. This looks like a spreadsheet, but you cannot change any values. Manually changing data is frowned upon in data science. Sound data science separates the raw data from the script and the outputs to ensure the process is repeatable.

Each row in the data for this case study results from a measurement. The data in this case study has the following fields (columns):

- Sample_No: Reference number of the sample

- Date_Sampled: The sampling date

- Sample_Point: The reference number of the sample point

- Suburb: The town in Gormsey

- Measure: The type of measurement

- Result: The result of the laboratory test

- Units: The units of the result (NTU, mg/l, or orgs/100ml)

3.3.2 Reading Tibbles

The readr package of the Tidyverse provides an alternative version to read CSV files and other rectangular data formats (Wickham et al., 2022c). The read_csv() function creates a tibble instead of a data frame. Hadley Wickham, the inventor of the Tidyverse, is from New Zealand, and tibble is a play on words on how somebody with a strong Kiwi accent would say table.

Tibbles are a bit faster when working with large data sets. They are also better at recognising the variable type in your file. In all other aspects, Tibble and data frames are precisely the same. Therefore, the words tibble and data frame are used interchangeably in the remainder of this book.

```
library(readr)
labdata <- read_csv("data/water_quality.csv")
```

The read_csv() function assumes that the first row contains the field names and the following rows contain the data, organised in columns. This function has many options you can use to fine-tune how R reads the data. The most common option is skip = n, which instructs the function to ignore the first n rows. This is useful because many CSV files contain metadata in the first rows. Chapter 7 discusses how to deal with unruly data. The readr package also has other functions to read rectangular data, which are all similar. Read the vignette to learn more about this package with vignette("readr").

3.3.3 Reading Excel Spreadsheets

Excel is by far the most popular software for storing and analysing data. While it is not an ideal tool to store data, the reality is that you will often need to read data in this format (Broman & Woo, 2018). The Tidyverse also contains the *readxl* package to read Excel spreadsheets (Wickham & Bryan, 2022). You will need to install this package separately. The code below initialises the library and reads the data worksheet, skipping the first two rows.

```
library(readxl)
labdata <- read_excel("data/water_quality.xlsx", skip = 2, sheet = "data")
```

Data in Excel is not always stored in an ideal format. The `read_excel()` function has a lot of options to control which cells you want to read from the spreadsheet. Use the help file to review the options or read the vignette to find out more details on how to wrangle spreadsheet data into R (`vignette("sheet-geometry")`).

3.4 Exploring Data Frames

A data frame has variables (columns) and observations (rows), also called rectangular data. The top row contains the variable names and the remainder of the values. So effectively, a data frame is a collection of vectors. Each variable needs to have the same number of observations, and all observations within a variable must be of the same data type, e.g., dates, text, and numbers.

The `names()` function lists the names of the variables. The `dim()` function results in a vector with the number of rows and columns. The `nrow()` and `ncol()` functions list the number of rows and columns for a data frame. The result of each function call is a single number.

```
names(labdata)

[1] "Sample_No"    "Date"        "Sample_Point" "Suburb"       "Measure"
[6] "Result"       "Units"

dim(labdata)

[1] 2422    7
```

When you run the name of a data frame in the console, R will show the first 1,000 lines, which is not very useful. You can use the `head()` and `tail()` functions to only show the first or last six rows. The `tibble` package in the Tidyverse is easier to use as it summarises the content. Any columns that don't fit on the screen are listed below the table.

```
labdata

# A tibble: 2,422 × 7
   Sample_No Date       Sample_Point Suburb    Measure        Result Units
       <dbl> <date>     <chr>        <chr>     <chr>           <dbl> <chr>
 1    603998 2069-01-01 ME_15385     Merton    Chlorine Total  0.18  mg/L
 2    603431 2069-01-01 ME_15385     Merton    E. coli         0     Orgs/100mL
 3    638433 2069-01-01 ME_12236     Merton    Chlorine Total  0.59  mg/L
 4    617355 2069-01-01 ME_12236     Merton    E. coli         0     Orgs/100mL
 5    663362 2069-01-03 SN_11009     Hallburgh Chlorine Total  0.08  mg/L
 6    618816 2069-01-03 SN_11009     Hallburgh Turbidity       0.2   NTU
 7    620121 2069-01-03 SN_11009     Hallburgh E. coli         0     Orgs/100mL
 8    627981 2069-01-03 ME_15385     Merton    E. coli         0     Orgs/100mL
 9    618060 2069-01-03 ME_15385     Merton    Turbidity       0.2   NTU
10    665782 2069-01-03 ME_15385     Merton    Chlorine Total  0.025 mg/L
# ... with 2,412 more rows
# i Use `print(n = ...)` to see more rows
```

3.5 Variable Types

The `dplyr` package in the Tidyverse provides the `glimpse()` function, which summarises the table variables, variables types, and the content.

```
glimpse(labdata)
```

```
Rows: 2,422
Columns: 7
$ Sample_No    <dbl> 603998, 603431, 638433, 617355, 663362, 618816, 620121, 6...
$ Date         <date> 2069-01-01, 2069-01-01, 2069-01-01, 2069-01-01, 2069-01-...
$ Sample_Point <chr> "ME_15385", "ME_15385", "ME_12236", "ME_12236", "SN_11009...
$ Suburb       <chr> "Merton", "Merton", "Merton", "Merton", "Hallburgh", "Hal...
$ Measure      <chr> "Chlorine Total", "E. coli", "Chlorine Total", "E. coli",...
$ Result       <dbl> 0.180, 0.000, 0.590, 0.000, 0.080, 0.200, 0.000, 0.000, 0...
$ Units        <chr> "mg/L", "Orgs/100mL", "mg/L", "Orgs/100mL", "mg/L", "NTU"...
```

Each line starts with the variable name and the variable type between <>. The Gormsey data has three types:

- `dbl`: Numeric values (double-precision floating-point format)

- `dttm`: Date / time variables (discussed in detail in Chapter 11)

- `chr`: Character variables

The `glimpse()` function is a specialised version of the `str()` function, which is part of the R base package.

R has several other variable types, such as factors for categorical data, logical variables (TRUE and FALSE), integers, and complex numbers. However, all variables must be of the same type in a vector. That is why all columns must hold data from one type in a data frame.

3.6 Explore the Content of a Data Frame

To view any variables within a data frame, you need to add the column name after a dollar sign: `labdata$Result`. When you execute this command, R shows a vector of the observations within this variable. For example, when you type a $ after `labdata`, RStudio will display a list of the available variables. You can pick the one you need with the arrow buttons and the enter key.

If you want to use only a subset of a vector, you can indicate the index number between square brackets. For example: `labdata$Results[1:10]` shows the first ten results, or `labdata$Results[c(1, 3, 5)]` to only show three values at positions 1, 3, and 5.

The most basic approach to subset a data frame is to add the number of rows and columns between square brackets. For a vector, you use one number, and for a data frame, two numbers: `[rows, columns]`. Note that `labdata$Date` results in a vector, and `labdata[, 2]` results in a new data frame.

For example, `labdata[1:10, 4:5]` shows the first ten rows and the fourth and fifth variables. R shows all values when there is no value in either the place for the rows or the columns. R is a mathematical language, and the index numbers thus start at one. In generic programming languages, the index starts at zero.

This syntax can also include the names of variables, e.g., `labdata[1:10, c("Suburb", "Result")]` shows the first ten rows of the Suburb and the Result variables. Besides numerical values, you can also add formulas as indices, for example `labdata[n * 2, ]`.

3.7 Filtering Data Frames

A data frame usually contains much more data than you need for analysis. Filtering is a basic activity to extract those parts that you need. You can filter the data using conditions. If, for example, you like to see only the turbidity data, then you can subset the data:

```
labdata[labdata$Measure == "Turbidity", ]
```

```
# A tibble: 734 × 7
   Sample_No Date       Sample_Point Suburb      Measure   Result Units
       <dbl> <date>     <chr>        <chr>       <chr>      <dbl> <chr>
 1    618816 2069-01-03 SN_11009     Hallburgh   Turbidity    0.2 NTU
 2    618060 2069-01-03 ME_15385     Merton      Turbidity    0.2 NTU
 3    617081 2069-01-03 SO_13252     Southwold   Turbidity    0.1 NTU
 4    679923 2069-01-03 TA_18683     Tarnstead   Turbidity    0.2 NTU
 5    635529 2069-01-03 SW_14121     Swadlincote Turbidity    0.2 NTU
 6    673777 2069-01-03 WA_10539     Wakefield   Turbidity    0.1 NTU
 7    676430 2069-01-05 BL_15407     Blancathey  Turbidity   2.24 NTU
 8    653828 2069-01-09 ME_12236     Merton      Turbidity    0.2 NTU
 9    666493 2069-01-09 WA_17036     Wakefield   Turbidity    0.2 NTU
10    683496 2069-01-09 SN_19642     Hallburgh   Turbidity    0.2 NTU
# ... with 724 more rows
# i Use `print(n = ...)` to see more rows
```

3.7.1 Logical Values

This method looks similar to what we discussed above. The difference is that the row indicator now shows an equation. When you execute the equation between brackets separately, you see a list of values that are either TRUE or FALSE. These values indicate whether the variable at that location meets the condition.

```
a <- 1:2
a == 1
```

```
[1] TRUE FALSE
```

Variables that are either TRUE or FALSE are called Boolean or logical. They are the building blocks of computer science. These variables indicate conditions and can also be used in calculations. The code below results in a vector with the values 2 and 0.

```
a <- c(TRUE, FALSE)
a * 2
```

```
[1] 2 0
```

You can use the common relational operators to test for complex conditions:

- p == q equal to each other

- p != q not equal to each other

- p < q less than

- p > q greater than

- p <= q less than or equal to

- p >= q greater than or equal to

R also evaluates relations between character strings using alphabetical order. In R, "Small" > "Large" results in TRUE because R compares strings lexicographically. You can build elaborate conditionals by combining more than one condition with logical operations. Table 3.1 shows the possible outcomes for the four most used logical operators.

TABLE 3.1 Truth table for basic logic operators.

p	q	!p	p & q	p | q	xor(p, q)
TRUE	TRUE	FALSE	TRUE	TRUE	FALSE
TRUE	FALSE	FALSE	FALSE	TRUE	TRUE
FALSE	TRUE	TRUE	FALSE	TRUE	TRUE
FALSE	FALSE	TRUE	FALSE	FALSE	FALSE

An example of a logical operation would be the expression "small" > "large" & 1 > 2, which results in FALSE because the first condition is true, but the second one is false, so they are not both true.

3.7.2 Filtering in Tidyverse

The Tidyverse method uses the filter() function from the dplyr package, which is more convenient than square brackets. The first parameter in this function is the data frame you need to filter. The second parameter is the condition. This code results in a new data frame with only turbidity measurements:

```
turbidity <- filter(labdata, Measure == "Turbidity")
```

This method is tidier than the brackets method because we don't have to repeat the data frame name and add a $ to the variables. We can apply this knowledge to the case study. The code below shows the THM samples in Tarnstead with a result greater than 0.1 mg/l. Note that testing for equality requires two equal signs.

```
filter(labdata, Suburb == "Tarnstead" & Measure == "THM" & Result > .1)
```

Now we can answer questions like how many turbidity results in all suburbs, except Strathmore, are lower than 0.1 NTU? To answer this question, subset the data for all results less than 0.1 NTU *and* where the Suburb is not Strathmore. The nrow() function counts the results.

```
nrow(filter(labdata, Suburb != "Strathmore" &
                    Measure == "Turbidity" &
                    Result < 0.1))
```

[1] 65

3.8 Counting Data

The last exploratory activity discussed in this chapter counts variables. Base R provides various methods to count entries in vectors and data frames. The `length()` function determines the length of a vector so that `length(labdata$Measure)` results in the number of rows in the data. The `unique()` function takes a vector and only shows the distinct parts.

```
length(labdata$Measure)
```

```
[1] 2422
```

```
unique(labdata$Measure)
```

```
[1] "Chlorine Total" "E. coli"        "Turbidity"      "THM"
```

The `table()` function in base R lets you quickly view the content of a data frame or a vector by counting each unique value. The `table()` function can take more than one variable to count one variable by another.

```
table(labdata$Measure)
```

```
Chlorine Total        E. coli           THM      Turbidity
           760            760           168            734
```

```
table(labdata$Suburb, labdata$Measure)
```

```
            Chlorine Total E. coli THM Turbidity
Blancathey             104     104  24       104
Hallburgh              105     105  24       105
Merton                 107     107  24       105
Southwold              105     105  24       105
Swadlincote            105     105  24       105
Tarnstead              105     105  24       105
Wakefield              129     129  24       105
```

The table output of this function is not a data frame, meaning it needs to be converted before we can further analyse it. The Tidyverse `dplyr` package provides a convenient alternative with the `count()` function. The result of this function is a new data frame, which makes it easy to use in subsequent steps. By default, the result of the count is stored in a new variable `n`.

```
count(labdata, Suburb, Measure)
```

```
# A tibble: 28 × 3
  Suburb     Measure            n
  <chr>      <chr>          <int>
1 Blancathey Chlorine Total   104
2 Blancathey E. coli          104
3 Blancathey THM               24
4 Blancathey Turbidity        104
5 Hallburgh  Chlorine Total   105
```

```
 6 Hallburgh   E. coli            105
 7 Hallburgh   THM                 24
 8 Hallburgh   Turbidity          105
 9 Merton      Chlorine Total     107
10 Merton      E. coli            107
# ... with 18 more rows
# i Use `print(n = ...)` to see more rows
```

Note that the output of this function is in long format instead of the wider format from the table() function. Chapter 7 discusses transforming data frames from wide to long formats (pivoting).

We can now combine these functions to create a table of the number of turbidity results for each Suburb. First, we create a subset of the data and then count the results. The output of this function is a new data frame with the count results. The name parameter declares a name for the variable that contains the counts. Note that when we create the new variable Samples, the name is between quotation marks. This syntax is required because R will otherwise look for a variable named Samples, which does not yet exist.

```
turbidity <- filter(labdata, Measure == "Turbidity")
turbidity_count <- count(turbidity, Suburb, name = "Samples")
turbidity_count
```

```
# A tibble: 7 × 2
  Suburb        Samples
  <chr>           <int>
1 Blancathey        104
2 Hallburgh         105
3 Merton            105
4 Southwold         105
5 Swadlincote       105
6 Tarnstead         105
7 Wakefield         105
```

The *dplyr* package includes the arrange() function if you like to sort the resulting data frame by one of its variables. Sorting is ascending by default and the helper function desc() creates a descending sort. If you only like to include the top 3 or top 10 in the results, then the slice_min() or slice_max() functions can be useful.

```
sample_count <- count(labdata, Sample_Point)
arrange(sample_count, n)
arrange(sample_count, desc(n))
slice_max(sample_count, n = 3, order_by = n)
```

3.9 Further Study

The most significant hurdle to learning any programming language is remembering the large volume of syntax and vocabulary. The Tidyverse website contains extensive documentation, including a series of cheat sheets. These are PDF files that succinctly show all functions in each package. However, all information on the Tidyverse website is part of the package library, so using help files and vignettes will give you all the information you need.

If you need some more advanced filtering of text data fields, then you can to use the *stringr* package, which assists in manipulating text variables (Wickham, 2022b). This package is loaded by default when enabling the Tidyverse libraries.

Almost all programming languages, including R, support Regular Expressions (often abbreviated as regex), which is a powerful tool to search in character strings (Fitzgerald, 2012). Regular expressions are like wildcards used in files, such as `*.R` for all files with the `.R` extension. Regular Expressions are, however, much more powerful.

If, for example, you want to filter the Gormsey data for all suburbs that start with the letter *M*, use the following line:

```
filter(labdata, str_detect(Suburb, "^M"))
```

```
# A tibble: 343 × 7
   Sample_No Date       Sample_Point Suburb Measure        Result Units
       <dbl> <date>     <chr>        <chr>  <chr>           <dbl> <chr>
 1    603998 2069-01-01 ME_15385     Merton Chlorine Total  0.18  mg/L
 2    603431 2069-01-01 ME_15385     Merton E. coli         0     Orgs/100mL
 3    638433 2069-01-01 ME_12236     Merton Chlorine Total  0.59  mg/L
 4    617355 2069-01-01 ME_12236     Merton E. coli         0     Orgs/100mL
 5    627981 2069-01-03 ME_15385     Merton E. coli         0     Orgs/100mL
 6    618060 2069-01-03 ME_15385     Merton Turbidity       0.2   NTU
 7    665782 2069-01-03 ME_15385     Merton Chlorine Total  0.025 mg/L
 8    625061 2069-01-09 ME_12236     Merton E. coli         0     Orgs/100mL
 9    653828 2069-01-09 ME_12236     Merton Turbidity       0.2   NTU
10    604923 2069-01-09 ME_12236     Merton Chlorine Total  0.05  mg/L
# ... with 333 more rows
# i Use `print(n = ...)` to see more rows
```

The `str_detect()` function matches the pattern in the `Suburb` variable. The caret symbol at the start indicates the first character. If you want to search for suburbs that end with an "n", then use "n$" as a filter. The *stringr* package has extensive documentation on the topic, which you can access with `vignette("regular-expressions")`.

Now that you can load data from a spreadsheet or CSV file, your next step will be exploring the content. Exploratory analysis is an essential part of the data science workflow. You can examine data graphically for some instant gratification, or you can look at the numbers for more precise answers. The next chapter explains using base R statistical functionality and the Tidyverse to explore data frames numerically using descriptive statistics to determine whether the results comply with the relevant regulations.

4

Descriptive Statistics

An essential skill of a data scientist is the ability to tell a story with data. The first step in telling these stories is summarising the available data. Descriptive statistics are tools that use a single number to describe many. Analysts calculate averages, medians, percentiles, and other statistical summaries to describe a data set's tendency, spread, positions, and shape. Statisticians have defined five types of descriptive statistics. We have already seen *measures of frequency* in Section 3.8, which count the number of times each observation occurs in the sample. The remaining four descriptive statistics are:

1. *Central tendency* summarises a distribution with a single number
2. *Position* describes how an observation relates to the others
3. *Dispersion* describes how far the data deviates from the central tendency
4. *Shape* summarises the shape of a distribution

Analysing water quality in a supply network starts with calculating the descriptive statistics. Water quality regulators worldwide have specified limits for many biological and chemical requirements that drinking water must comply with. They set minimum and maximum values and percentiles for water quality measures to ensure that the water is safe to drink and does not taste or smell nasty. Descriptive statistics describe samples of data. In water quality, the word sample has a different meaning. A sample is a small amount of water representing the body of water we want to analyse. From this sample, one or more measurements are taken. In statistics, a sample is a series of numbers randomly taken from a population. So confusingly, collecting water quality sample results is a mathematical sample. One mathematical sample thus consists of the measurement results of multiple physical samples. This chapter uses the word sample in the statistical sense of the word (McBride, 2005).

This chapter shows how to calculate descriptive statistics using the water quality data from the first case study. The learning objectives for this chapter are:

- Summarise water quality data using descriptive statistics
- Evaluate compliance of water quality data with relevant regulations
- Perform grouped data analysis

4.1 Case Study Problem Statement

The case study for this chapter is the Gormsey water quality data used in the previous chapter. The water quality regulations for the fictional city of Gormsey specify the following requirements:

- *Escherichia coli*: All collected drinking water samples contain no E. coli per 100 millilitres of drinking water, except false positive samples

- *Total Chlorine*: Not less than 1 mg/l and not more than 5 mg/l

- *Trihalomethanes*: Less than or equal to 0.25 mg/l of drinking water

- *Turbidity*: The 95th percentile of results for samples in any 12 months must be less than or equal to 5.0 Nephelometric Turbidity Units

The data and code for this chapter are available in the data folder of your RStudio project (see Section 2.3.1). The examples in this chapter focus on turbidity measurements. Before we get started, we first need to load the data and extract the turbidity measurements.

```
library(readr)
library(dplyr)
labdata <- read_csv("data/water_quality.csv")
turbidity <- filter(labdata, Measure == "Turbidity")
```

4.2 Measures of Central Tendency

The central tendency, or expected value, of a series of measurements, is a single value that describes the sample with one number. The most common central tendency measures are the mean, median, and mode (Figure 4.1).

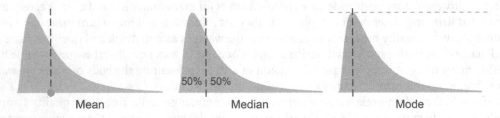

FIGURE 4.1 Measures of central tendency.

4.2.1 Mean

The mean() function calculates the arithmetic mean of a vector of numbers $\bar{x}$ (the sample mean, pronounced x-bar), which is the sum of all values in the sample, divided by the number of values. Just for completeness, the code below also shows how to calculate indirectly. For convenience and brevity, the turbidity results are stored in the variable x and the number of observations in the variable n.

```
x <- turbidity$Result
n <- length(x)

sum(x) / n

[1] 0.360327

mean(x)

[1] 0.360327
```

R also has a function to calculate the weighted mean, where each observation is multiplied by a factor, for example, frequency tables. The `weighted.mean()` function has two parameters, the observations and their weights. In the example below, we have 7 samples at 0.2 NTU, 5 samples of 0.5 NTU, and 2 samples of 1 NTU. The weighted mean, in this case, is the same as when we have a data set where the values are repeated. The `rep()` function replicates to create a sequence.

```
weighted.mean(c(0.2, 0.5, 1), w = c(7, 5, 2))
```

```
[1] 0.4214286
```

```
mean(rep(c(0.2, 0.5, 1), c(7, 5, 2)))
```

```
[1] 0.4214286
```

4.2.2 Median

The median of a sample is the value where half the sample is lower, and half the sample is higher. To find the mean, you first rank all observations in order and then take the middle one (or interpolate when there are an uneven number of observations). The median is the same as the 50th percentile (Section 4.3).

```
median(x)
```

```
[1] 0.2
```

In a symmetric distribution, the mean and the median are the same. The difference between the mean and median thus tells you something about the shape of the distribution. Therefore, the median tends to be more appropriate as a central tendency measure for asymmetric distributions.

4.2.3 Mode

The third common central tendency is the mode, which is the value that appears most often in the data. Unfortunately, r has no standard function for determining the mode, so we create our own function, demonstrating the extensibility of the language. Functions are discussed in more detail in Chapter 12. Calculating the mode is tricky when the variables have high accuracy, as there will not be many duplicates, so you might have to round the numbers discussed in Section 6.7.

```
mode <- function(x) {
  ux <- unique(x)
  ux[which.max(tabulate(match(x, ux)))]
}
```

```
mode(x)
```

```
[1] 0.2
```

4.3 Measures of Position

A measure of position is a number's relative position within the sample. Examples of measures of position are quartiles, deciles, and percentiles, generically known as quantiles. The first quartile is the value greater than one-quarter of the way through an ordered data set. Deciles are the same: they split the data into ten parts, and percentiles split the data into one hundred parts. The median is the same as the fifth decile, second quartile, or fiftieth percentile. Percentiles are a standard method to describe water quality data. For example, if we state that the 95th percentile of turbidity was 4 NTU, then 95% of results were lower than or equal to 4 NTU, allowing the occasional spike.

You can display a five-number summary of a vector with the summary() function, which provides the minimum, two quartiles, median, and maximum values as percentiles.

```
summary(x)

  Min. 1st Qu.  Median   Mean 3rd Qu.    Max.
0.0500  0.1000  0.2000  0.3603  0.2000  8.8000
```

The quantile() function calculates the quantiles of a vector of numbers. A quantile is simply a percentile divided by 100. By default, this function provides the same five values as the summary function. Confusingly, the output of the quantile function is displayed as percentiles.

```
quantile(x)

  0%   25%   50%   75%  100%
0.05  0.10  0.20  0.20  8.80
```

The probs parameter lets you define which quantiles you like to see. For example, quantile(x, probs = 0.95) provides the 95th percentile of the turbidity measurements. Note that the probs parameter is a quantile between zero and one. The quantile function can also take a vector of one or more probabilities to calculate different outcomes. For example quantile(x, c(0.33, 0.66)) results in a vector with two variables. You can omit the probs part because it is the default parameter on the second position. Parameter names can be omitted in an R function when you enter the value in the same order as shown in the help file.

4.3.1 Calculating Percentiles

The water quality regulator in Gormsey specifies that turbidity percentiles should be calculated with the 'Weibull Method'. What does this mean, and how can R help you comply with this regulation? The help text for the quantile() function mentions nine methods to calculate percentiles and refers to a journal article for details (Hyndman & Fan, 1996), so let's investigate further. Calculating a percentile requires three steps, illustrated in Figure 4.2 (McBride, 2005):

1. Place the observations in ascending order: $y_1, y_2, \ldots x_n$

2. Calculate the rank (r) of the required percentile

3. Interpolate between adjacent numbers

Where:

- y: Observation

- r: Ranking number

- $\lfloor r \rfloor$: Floor or r (round a value to its nearest lower integer)

- $\lceil r \rceil$: Ceiling of r (round a value to its nearest higher integer)

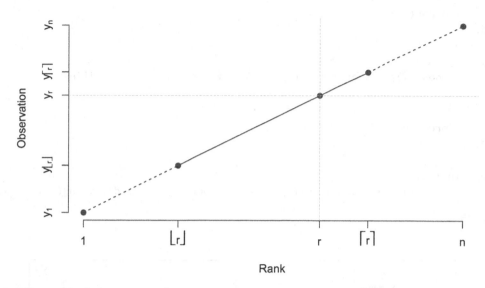

FIGURE 4.2 Calculating percentiles.

With 52 ranked weekly turbidity samples, the 95th percentile lies between sample 49 and 50: $r = pn = 0.95 \times 52 = 49.4$. To determine the percentile, you interpolate between the measured values. If, for example, sample 49 and 50 ($y_{\lfloor r \rfloor}$ and $y_{r\lceil r \rceil}$) are 0.5 and 1.0 NTU, then the 95th percentile y_r is:

$$y_r = y_{\lfloor r \rfloor} + (y_{r\lceil r \rceil} - y_{\lfloor r \rfloor})(r - \lfloor r \rfloor) \tag{4.1}$$

$$y_r = 0.5 + (1 - 0.5) \times (49.4 - 49) = 0.7$$

However, the method where $r = pn$ is only valid for normally-distributed samples. Therefore, statisticians have defined several methods to determine percentiles, each of which is most suitable for specific situations. These approaches all use different rules to determine the rank r.

Hyndman & Fan (1996) give a detailed overview of each of the nine methods. This paper provides the Weibull method with the less poetic name of $\hat{Q}_6(p)$ because it is the sixth option in their list. Waloddi Weibull, a Swedish engineer famous for his statistical distribution, was one of the first to describe this method. In his method, the rank r of a percentile p is given by:

$$r = p(n + 1) \tag{4.2}$$

For a sample of 52 turbidity tests, the 95th percentile thus lies between ranked result number 50 and 51: $r = 0.95(52 + 1) = 50.35$. This method gives a higher result than the standard method for normal distributions. The Weibull method is more suitable for positively skewed distributions, as is often the case with water quality. Laboratory data generally has a lot of low values, with the occasional spikes at high values. Please note that there is no one correct method to calculate percentiles. The most suitable method depends on the population distribution and the analysis's purpose. In this case study, the regulator prescribes the Weibull method, which is biased towards the high end of the distribution.

The script below compares the Weibull and the linear methods (Figure 4.3). The code orders the samples from low to high and sets the probability and length parameters. The ranks for the Weibull and linear methods have different formulas, after which we can calculate the percentiles. With highly skewed data, as is often the case with turbidity measurements, the Weibull method results in a higher percentile value.

```
x_ord <- x[order(x)]
p <- 0.95

r <- p * n
(1 - (r - floor(r))) * x_ord[floor(r)] + (r - floor(r)) * x_ord[ceiling(r)]

[1] 0.4

x_ord <- x[order(x)]
p <- 0.95

r <- p * (n + 1)
(1 - (r - floor(r))) * x_ord[floor(r)] + (r - floor(r)) * x_ord[ceiling(r)]

[1] 0.425
```

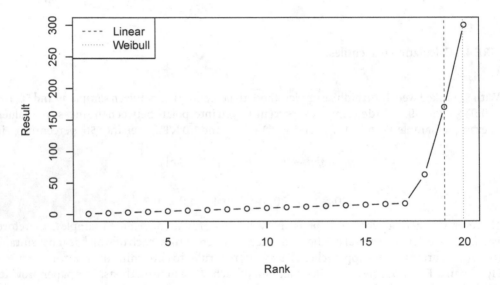

FIGURE 4.3 Comparing calculation types for percentiles.

The `type` parameter in the quantile function defines which of the nine methods is used. Type 7 is the default method, and the Weibull method is type 6. The linear interpolation method in Figure 4.2 is type 4. The `fivenum()` function provides a similar summary without the mean value and uses type 2.

```
quantile(x, 0.95,  type = 4)

95%
0.4

quantile(x, 0.95,  type = 6)

  95%
0.425
```

In conclusion, the Weibull method results in a slightly higher percentile than the default approach because of the particular shape of the distribution (Section 4.5). The regulator chooses this method because it is more conservative as it trends closer to the higher values.

4.4 Measures of Dispersion

A single number is never sufficient to describe a distribution of a sample of values. For example, an average value of 0.2 NTU could still mean that we have regular spikes several times that number. Dispersion is the extent to which the measurements are spread. Several measures are available to describe dispersion.

4.4.1 Range

As the name suggests, the range of a sample is defined by its lowest and highest values. R has three functions to inspect the range of a vector. The `diff()` function calculates the difference between two consecutive values in a vector.

```
min(x)
max(x)
range(x)
diff(range(x))

[1] 0.05
[1] 8.8
[1] 0.05 8.80
[1] 8.75
```

4.4.2 Inter-Quartile Range

A data set's Inter-Quartile Range (IQR) is the difference between the third and first quartile. It measures the gap between the central values when the data is arranged in ascending order. We can calculate it with the `quantile()` function. The `IQR()` function is a convenient shortcut to calculate this value. This measure provides similar information to the range of the data, but it is less susceptible to outliers, as shown in the boxplot (Section 4.7.2).

```
diff(quantile(x, probs = c(0.25, 0.75)))

75%
0.1

IQR(x)

[1] 0.1
```

4.4.3 Variance

The variance of a population indicates how far the data deviates from the mean. A low variance suggests that the sample values are close to the mean. The variance of a population σ^2 is the average squared difference between the observations x and the population mean μ. The differences are squared because otherwise, the sum would be zero (Equation 4.3).

$$s^2 = \frac{1}{n-1} \sum_{i=1}^{n} (x_i - \bar{x})^2 \tag{4.3}$$

We don't know the population distribution of possible measurements (the whole body of water), as we only have a sample. We need to subtract one from the number of samples n (Bessel's correction) to estimate the unbiased sample variance (s^2).

Calculating the variance of a sample means you only have a fraction of all possible values, so the answer will be biased, which means that it has an error. A statistical error is not a mistake but a deviation from the actual value. Any value in a sample will be closer to the sample mean than the population mean. Subtracting one from the sample size corrects this error. The var() function in R calculates the sample variance. The second line of code calculates the sample variance from the first principles, demonstrating the internal workings of this function.

```
var(x)
```

```
[1] 1.050882
```

```
sum((x - mean(x))^2) / (n - 1)
```

```
[1] 1.050882
```

4.4.4 Standard Deviation

A problem with using variance to describe results is that it is not in the same units as the measurement. Standard deviation s is the square root of a sample's variance and is in the same units as the measurement. The sd() formula calculates the standard deviation of a sample.

```
sd(x)
```

```
[1] 1.025126
```

```
sqrt(sum((x - mean(x))^2) / (n - 1))
```

```
[1] 1.025126
```

4.5 Measures of Shape

Lastly, we also need to know something about the shape of the distribution. Although we can review the shape of the distribution with a histogram, it is good to be able to express the shape in numbers. Two methods are available. Skewness measures the symmetry of the distribution and kurtosis whether the tails are fat or narrow.

4.5.1 Skewness

A standard normal distribution is a nice symmetrical bell curve where the mean and median have the same value. In a skewed distribution, most observations are either low or high (Figure 4.4). Water quality data is typically positively skewed, which means that most values are small. The long tail of the distribution represents outliers so a very high or low skewness value can indicate the presence of anomalies.

Statisticians have developed several methods to calculate the skewness of a distribution, also called the coefficient of asymmetry. The most common method uses the third and second moments of the distribution. We have already seen the second moment in variance calculations. The Fisher-Pearson coefficient of skewness is defined as the standardised third moment γ_1, which is the third moment divided by the standard deviation to the third power:

$$\gamma_1 = \frac{\frac{1}{n}\sum_{i=1}^{n}(x-\bar{x})^3}{\sigma^3} \tag{4.4}$$

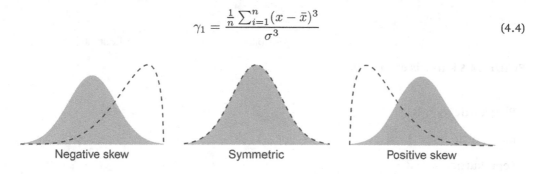

Negative skew   Symmetric   Positive skew

FIGURE 4.4 Skewness examples.

We can calculate the normalised third central moment from first principles (Equation 4.4).

```
(sum((x - mean(x))^3) / n) / (sqrt(sum((x - mean(x))^2) / n)^3)
```

```
[1] 5.663652
```

This is a cumbersome way to calculate skewness. The moments package provides the skewness() formula to simplify these calculations (Komsta & Novomestky, 2022). Note that the formula in the *moments* package calculates the population skewness without a correction for sample bias. For large sample numbers, the difference will be negligible.

```
moments::skewness(x)
```

```
[1] 5.663652
```

We find that the skewness of this distribution is positive, which means that there are more low than high values. The magnitude of skewness has no relationship to the observations. The turbidity data is highly skewed, following this rule of thumb:

- *Approximately symmetric*: Between -½ and +½

- *Moderately skewed*: between -1 and -½ or between +½ and +1

- *Highly skewed*: less than -1 or greater than +1

4.5.2 Kurtosis

The kurtosis of a sample measures how heavily the tails of a distribution differ from the tails of a normal distribution. In other words, the kurtosis shows whether the tails of a sample contain extreme values. Samples with small kurtosis are 'fatter' than those with high kurtosis and have a higher likelihood of outliers.

A normal distribution has a kurtosis of three and is called mesokurtic, which is a fancy way of saying medium (*meso*). Distributions with a kurtosis higher than three are called leptokurtic, meaning they have a thinner (*lepton*) central part. Conversely, platykurtic distributions have a flatter (*platys*) central part and a kurtosis less than three (Figure 4.5). In some definitions, the kurtosis is normalised by subtracting three so that a normal distribution has zero kurtosis.

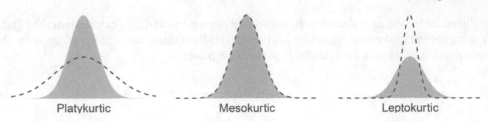

| | | |
| Platykurtic | Mesokurtic | Leptokurtic |

FIGURE 4.5 Kurtosis examples.

- Platykurtic $\gamma_2 < 3$
- Mesokurtic: $\gamma_2 = 3$
- Leptokurtic: $\gamma_2 > 3$

Kurtosis γ_1 can be measured with the standardised fourth moment of the distribution. Because the formula uses the fourth power, outliers contribute a lot to the measure of kurtosis (Equation 4.5).

$$\gamma_2 = \frac{\frac{1}{n}\sum_{i=1}^{n}(x - \bar{x})^4}{\sigma^4} \tag{4.5}$$

```
(sum((x - mean(x))^4) / n) / (sqrt(sum((x - mean(x))^2) / n)^4)
```

```
[1] 36.59041
```

The *moments* library spares you the hard work of writing the kurtosis formula from scratch:

```
moments::kurtosis(x)
```

```
[1] 36.59041
```

The turbidity data has a kurtosis much larger than three, so it is considerably leptokurtic.

4.6 Analysing Grouped Data

Now we can put all this statistical knowledge to good use. In the previous chapter, we created a filtered data set to determine statistics for a specific suburb and parameter. Doing this manually for each suburb or parameter would be a tedious task. As a general rule in coding, if you copy and paste code more than twice, there is a more efficient way of doing what you are trying to achieve. The group_by() function in the *dplyr* library simplifies analysing data from multiple samples by grouping data and analysing the results for each group (Wickham et al., 2022b).

Using group_by() splits the data frame into a group of smaller ones, filtered by the grouping variable. This example creates a new data frame with turbidity data grouped by Measure. When you view this data frame (tibble) in the console, it shows the normal output, but with an added comment that it is grouped by suburb, and that there are seven groups.

```
library(dplyr)
labdata_grouped <- group_by(labdata, Measure)
labdata_grouped
```

```
# A tibble: 2,422 × 7
# Groups:   Measure [4]
   Sample_No Date       Sample_Point Suburb    Measure        Result Units
       <dbl> <date>     <chr>        <chr>     <chr>           <dbl> <chr>
 1    603998 2069-01-01 ME_15385     Merton    Chlorine Total  0.18  mg/L
 2    603431 2069-01-01 ME_15385     Merton    E. coli         0     Orgs/100mL
 3    638433 2069-01-01 ME_12236     Merton    Chlorine Total  0.59  mg/L
 4    617355 2069-01-01 ME_12236     Merton    E. coli         0     Orgs/100mL
 5    663362 2069-01-03 SN_11009     Hallburgh Chlorine Total  0.08  mg/L
 6    618816 2069-01-03 SN_11009     Hallburgh Turbidity       0.2   NTU
 7    620121 2069-01-03 SN_11009     Hallburgh E. coli         0     Orgs/100mL
 8    627981 2069-01-03 ME_15385     Merton    E. coli         0     Orgs/100mL
 9    618060 2069-01-03 ME_15385     Merton    Turbidity       0.2   NTU
10    665782 2069-01-03 ME_15385     Merton    Chlorine Total  0.025 mg/L
# ... with 2,412 more rows
# i Use `print(n = ...)` to see more rows
```

We can use grouped data frames to compute summary statistics for each group using the `summarise()` function. For example, to calculate the average and maximum value of the result for each suburb. The code below provides results grouped by measure.

```
summarise(labdata_grouped,
          Minimum = min(Result),
          Median = median(Result),
          p95 = quantile(Result, 0.95, type = 6),
          Maximum = max(Result),
          Kurtosis = moments::kurtosis(Result))
```

```
# A tibble: 4 × 6
  Measure        Minimum Median   p95 Maximum Kurtosis
  <chr>            <dbl>  <dbl> <dbl>   <dbl>    <dbl>
1 Chlorine Total  0.025  0.33   1.54    2.04     2.97
2 E. coli         0      0      0       3      378.
3 THM             0.0005 0.0045 0.13    0.763   35.6
4 Turbidity       0.05   0.2    0.425   8.8     36.6
```

The `summarise()` function uses the grouped data frame and creates the new variables for each suburb. You can add any R function within the summarise statement to easily calculate any grouped statistic. Use the `n()` function to count the number of observations (rows) in each group. You can also group a data frame or tibble by more than one variable. Note that the example below results in the same output as `count(labdata, Measure, Suburb, name = "Samples")`.

```
labdata_grouped <- group_by(labdata, Measure, Suburb)
summarise(labdata_grouped,
          samples = n())
```

```
`summarise()` has grouped output by 'Measure'. You can override using the
`.groups` argument.
# A tibble: 28 × 3
# Groups:   Measure [4]
  Measure        Suburb          samples
  <chr>          <chr>             <int>
```

```
 1 Chlorine Total Blancathey      104
 2 Chlorine Total Hallburgh       105
 3 Chlorine Total Merton          107
 4 Chlorine Total Southwold       105
 5 Chlorine Total Swadlincote     105
 6 Chlorine Total Tarnstead       105
 7 Chlorine Total Wakefield       129
 8 E. coli        Blancathey      104
 9 E. coli        Hallburgh       105
10 E. coli        Merton          107
# ... with 18 more rows
# i Use `print(n = ...)` to see more rows
```

When grouping by more than one variable, the resulting data frame is also grouped. The un-group() function reverts it to a regular tibble.

4.7   Basic Data Visualisation

After looking at these cold hard numbers, visualising the results is a good idea. Visualisation is important because it allows the analyst to observe possible patterns that the summary statistics cannot communicate. Francis Anscombe (1973) devised four data sets, shown in Figure 4.6, that have the same mean values, sample variance, correlation, and regression line.

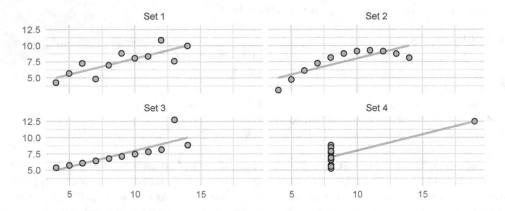

FIGURE 4.6 Anscombe's quartet.

However, visualising these sets emphasises the differences. Set two should not be modelled with a linear relationship. The third and fourth sets suffer from outliers in the observations that skew the results. You can explore this famous data set with the data() function, which makes built-in data sets accessible. Chapters 8 and 9 explore correlation and regression.

```
data(anscombe)
anscombe
```

4.7.1 Histograms

The most common visualisation to assist with descriptive statistics is the histogram. A histogram shows the frequency of groups of values in the sample as a bar chart. The `hist()` function displays a simple histogram, shown in Figure 4.7. Water quality data tends to be asymmetrical due to the predominantly low results with only a few high values. Therefore, it might be useful to transform the data so that it more closely resembles a normal distribution (see Section 9.5.2).

```
par(mfcol = c(1, 2), mar = c(4, 2, 1, 0))
hist(turbidity$Result, main = "No transformation")
hist(log(turbidity$Result), main = "Log transformation")
```

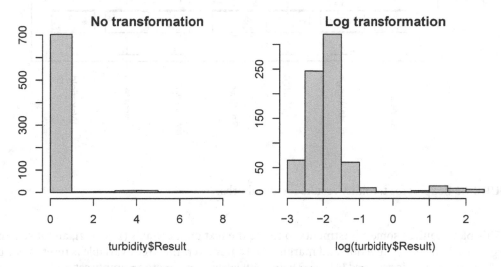

FIGURE 4.7 Histograms of turbidity data without and with log transformation.

The `par()` function splits the plotting screen into two columns with the `mfcol` parameter. The second parameter for a histogram is the location or number of breaks between the bars. By default, R uses an algorithm to define optimum intervals. The `breaks` parameter changes the interval

```
hist(log(turbidity$Result), breaks = 5)
```

4.7.2 Box and Whisker Plots

Another method to visualise a distribution is the Box and Whisker plot, also called boxplot (Figure 4.8). The `boxplot()` function conveniently summarises the distribution of observations. The grey box shows the Inter Quartile Range (IQR, Section 4.4.2), and the dark line is the median (Section 4.2.2). The whiskers extend to the first recorded value below 1.5 times the IQR from the 25th and above the 75th percentile, or the minimum and maximum, whichever is the greatest. Any values smaller or larger than the whiskers are marked as outliers with a circle. The `range` parameter defines the extent of the outliers, which is 1.5 by default. Chapter 12 discusses anomaly detection in detail.

This function can take a vector as input to draw a graph for one value, but you can also use a more complex syntax to compare values. The code below uses the `data` parameter to specify a data frame. The second parameter instructs R to plot the natural logarithm of the results by suburb. The tilde (~) describes relationships between variables.

```
par(mar = c(8, 4, 4, 1), mfcol = c(1, 1))
boxplot(data = turbidity, log10(Result) ~ Suburb, las = 2,
        xlab = NULL, main = "Gormsey Turbidity Samples")
```

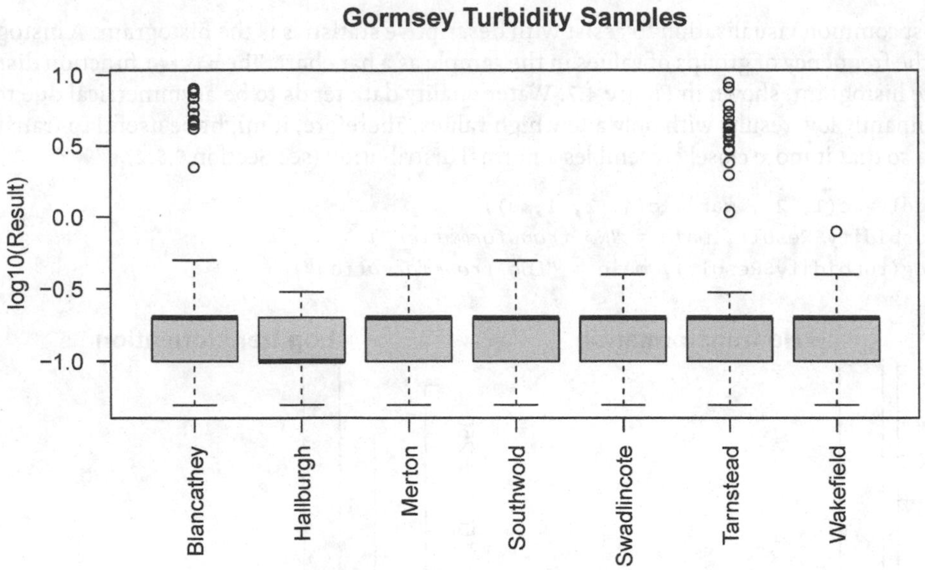

Gormsey Turbidity Samples

FIGURE 4.8 Boxplot for turbidity results for all suburbs.

This plot requires some adjustments to rotate the text on the x-axis (`las = 2`), and it sets the margins with the `par()` function and margin (`mar`) parameter. The `mfcol` variable is reset to one by one. The x-axis label is removed (`xlab = NULL`). A `NULL` value refers to an empty set.

4.8   Further Study

Many of the thousands of available R packages provide competing methods for the same computation. Some methods might be faster than others or have more flexible parameters to fine-tune the outcomes. For example, several packages are available to calculate skewness and kurtosis.

The skewness and kurtosis formulas in the *moments* package calculate the population statistic without correcting for sample bias. For large sample numbers, the difference is negligible. The *e1071* package provides functionality for machine learning, including skewness and kurtosis formulas that adjust for sample bias (Meyer et al., 2023). Using this method barely changes the outcome of this case study. The advanced formulas should be used for small sample sizes. The name of this package is the code for the department of statistics at TU Wien, where the code was developed.

This chapter showed how to calculate descriptive statistics for a single variable (univariate). However, when looking at more than one variable, we can use multivariate descriptive statistics, such as covariance and correlations, explained in Chapter 8.

The R language has good visualisation capabilities out-of-the-box, and you have seen some plotting functions in this and previous chapters. The graphs we created in this chapter are *exploratory*. They are made for the analyst as part of the data science workflow and to find the data story. The next step is creating *explanatory* graphs to communicate the data story. The next chapter discusses telling data stories with the *ggplot2* package.

5

Visualising Data with ggplot2

Advertising executive Fred Barnard coined the well-worn cliché that "a picture is worth a thousand words" in 1927. Perhaps we should say: "A graph is worth a thousand data points". A simple chart can summarise thousands of numbers and make them understandable in the blink of an eye. The Anscombe Quartet discussed in the previous chapter (Figure 4.6) demonstrates the importance of visualising data to uncover patterns that numeric analysis might not detect. Visualisation is essential for a data scientist who wants to explore and understand data and communicate results. The internet is awash with infographics and other creative methods to generate images from data. However, not all data visualisations are created equal. Some designs are hard to interpret, leading to wrong decisions or inaction. Others confuse or even deceive the reader.

Scientists have studied how the mind perceives graphics in great depth and devised a comprehensive body of knowledge that helps us create sound, useful, and aesthetic visualisations. The first part of this chapter introduces some principles of best practice in data visualisation. The second part presents the *ggplot2* library and some basic techniques to create high-quality graphics. The learning objectives for this chapter are:

- Evaluate data visualisations using basic principles

- Apply the principles of the *Grammar of Graphics*

- Visualise water quality data with the *ggplot2* package

5.1 Principles of Visualisation

Chapter 1 explained how the principles of good data visualisation align with the art of architecture. Although visualising data has some parallels with art, it is a different craft. All works of art are a form of deception. An artist paints a three-dimensional world on a flat canvas, and although we see people and buildings, we are looking at mere blobs of paint. Data visualisation is not art because it needs to be truthful and not deceive. The purpose of any graph is to reliably reflect the data without leaving room for the viewer to wrongly interpret the scene. Following some basic rules prevents confusing the consumers of your data products.

Firstly, visualisation needs to have a straightforward narrative. The reader should be able to draw the intended conclusion without mental effort. Data visualisation is not a *Where's Wally* where the viewer has to search for the point of interest. A good data visualisation tells a story, such as comparing numbers, showing a trend or anything else of interest. Charts, for the sake of themselves, are a waste of space and time. Many organisations maintain massive dashboards that contain everything but the kitchen sink. While these displays are impressive, they don't necessarily add value.

A visualisation should ideally contain a point of interest, such as the most recent, the highest, the lowest, comparison, or anything else worth noting. Secondly, visualisations should have a minimalist design. Remove elements that don't add to the story, such as overuse of colour or backgrounds or unnecessary third dimensions. Graphs with too many lines or colours confuse the reader and increase the difficulty of interpreting the information.

5.1.1   Aesthetics of Visualisation

Aesthetics in data science is not about creating works of art but about producing useful images. But having said this, there is a lot we can learn from art when thinking about data visualisation. Some visualisations remind me of the works by Jackson Pollock, famous for his chaotic abstract paintings. Engineers love to maximise the number of visual elements in a graph, with lines and colours splashed across the screen. Adding lots of graphical elements and colours to a chart reduces its usability. Perhaps a good data visualisation should look more like a painting by Piet Mondrian, famous for his austere compositions with straight lines and primary colours. Using art to explain data visualisation is not an accidental metaphor because visual art represents how the artist perceives reality. Therefore, this comparison between Pollock and Mondrian does not judge their artistic abilities. For Pollock, reality was chaotic and messy, while Mondrian saw a geometric order behind the perceived world. Data visualisation is an art in this sense, as it also attempts to express reality in an abstract image.

There are no formulas or algorithms to ensure perfect visualisations. The aesthetics of data visualisation is in the eye of the beholder. However, we can define data beauty with a simple heuristic when viewing aesthetics from a practical perspective. Edward Tufte is an American statistician famous for his visualisation work. Tufte introduced the concept of the data-to-ink ratio. In simple terms, this ratio expresses the relationship between the ink on the paper that tells a story and the total amount of ink used. Tufte (2001) argues that this ratio should be as close as possible to one. In other words, we should not use any graphical elements that don't communicate information, such as background images, redundant lines, and lengthy text. In the paperless era, we use the data-to-pixel ratio as a generic measure for the aesthetics of visualisations. Unnecessary lines, multiple colours or narratives risk confusing the report user. The data-to-pixel ratio is not a mathematical concept that needs to be expressed in exact numbers. Instead, this ratio is a rule of thumb for designers of visualisations to help them decide what to include and, more importantly, what to exclude from an image.

Figure 5.1 shows an example of a low and high data-to-pixel ratio. The bar chart on the left has a low ratio. The background image of a man drinking water might be on-topic, but it only distracts from the message. Using colours to identify the variables is unnecessary because the labels are at the

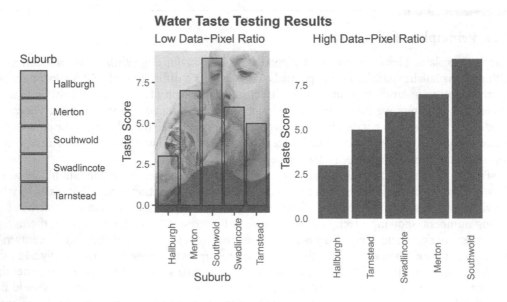

FIGURE 5.1 Examples of low and high data-to-pixel ratios.

bottom of the graph. The legend is not very functional because it also duplicates the labels. Lastly, the lines around the bars have no function.

In the improved version, all unnecessary graphical elements have been removed. Assuming that the story of this graph is to compare taste results, the columns have been ranked from large to small. Suppose the narrative of this graph was to compare one or more of the variables with other variables. In that case, groups of bars can be coloured to indicate the categories. The basic rule of visually communicating data is not to embellish your visualisations with unnecessary graphical elements or verbose text that does not add to the story. When visualising data, austerity is best-practice.

5.1.2 Telling Stories

Visualisations need to tell a story. The story in a graph should not be a mystery novel. A visualisation should not leave the viewer in suspense but get straight to the point. There is no need for spoiler alerts as we want to create clarity, not puzzles.

Trying to squeeze too much information into one graph confuses readers. Ideally, each visualisation should contain only one or two narratives. However, sometimes it is better to create multiple charts than to combine everything you want to visualise in one image.

Numerical data can contain several types of narratives. A graph can compare data points to show a trend among items or communicate differences. Bar charts are the best option for comparing data points with each other. A line graph is possibly your best option to compare data points over time. The distribution of data points is best visualised using a histogram or a boxplot. Scatter plots or bubble charts show relationships between two or three variables (Figure 5.2).

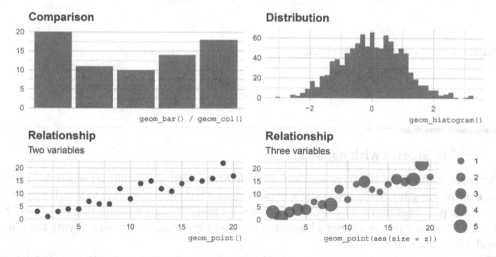

FIGURE 5.2 Examples of stories told with quantitative data.

Every visualisation needs to tell a story and not just summarise a bunch of numbers. The detailed considerations of choosing the most suitable visualisation are outside the scope of this book. However, the internet contains many tools to help you with this choice. Andrew Abela developed the Chart Chooser to select the most suitable visualisation.[1]

---

[1] Andrew Abela, Chart Chooser (extremepresentation.com).

5.1.3   Visualising Data in R

The R language has a base plotting functionality used in the previous chapters. This method starts with a blank canvas and adds elements one by one. Then, you create the main plot, add lines and labels, and so on. The base system has a layered approach with many tools to modify and annotate your visualisation.

Several packages, such as *lattice*, extend the basic functionality. However, the *ggplot2* package is R's most popular visualisation tool, as it uses sensible defaults and is easily extensible with themes and custom visualisations. This package is more aligned with principles of good visualisation design, and the remainder of the chapter shows how to tell data stories using *ggplot2*.

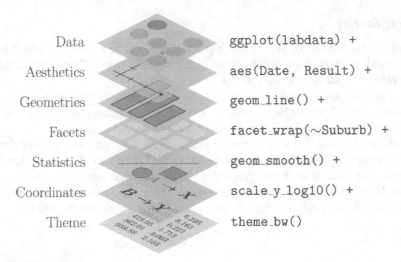

Data	`ggplot(labdata) +`
Aesthetics	`aes(Date, Result) +`
Geometries	`geom_line() +`
Facets	`facet_wrap(~Suburb) +`
Statistics	`geom_smooth() +`
Coordinates	`scale_y_log10() +`
Theme	`theme_bw()`

FIGURE 5.3 The Grammar of Graphics.

5.2   Telling Stories with *ggplot2*

In the previous chapter, you have seen some of the basic plotting capabilities of the R language. These base plotting functions are ideal for exploring data but not always easy to use. The Tidyverse collection of packages contains *ggplot2*, one of the most powerful data visualisation tools (Wickham et al., 2022a). This package follows a layered approach to visualising data, simplifying the process of producing sophisticated graphics.

This section introduces the basics of *ggplot2* using the Gormsey water quality data. The data and code for this chapter are available in the data folder of the RStudio project. Chapter 3 describes how to obtain this data. The *ggplot2* package applies the *Grammar of Graphics* developed by Leland Wilkinson (2005), hence the *gg* in ggplot. This grammar is an approach to systematically creating visualisations in logical layers data (Figure 5.3).

- *Data* exists at the lowest level, without which there is nothing to visualise.

- *Aesthetics* defines which graph variables are visualised and how they look (colour, line shapes, and sizes).

- *Geometries* are the shapes representing the data, such as bars, pies, or lines.

- *Facets* can be used to divide a visualisation into subplots.

- *Statistics* relate to any specific transformations to summarise the data, such as trend lines.

- *Coordinates* define how data is represented on the canvas. Primarily used in mapping.

- *Themes* define all the non-data pixels (font sizes, backgrounds, and so on).

The *ggplot2* package uses this vocabulary and layered approach to build visualisations. The following sections show examples using the Gormsey water quality data, building complexity layer by layer.

5.2.1 Data Layer

The ggplot() function always starts with the data variable, which has to be a data frame or a Tibble. You can use scalars and vectors in the base plotting functions, but *ggplot2* only accepts data frames. So, if you evaluate this function with only a data frame, it will draw a canvas and nothing else (left image in Figure 5.4). This is the canvas on which we build all further layers.

```
library(readr)
labdata <- read_csv("data/water_quality.csv")

library(ggplot2)
ggplot(labdata)
```

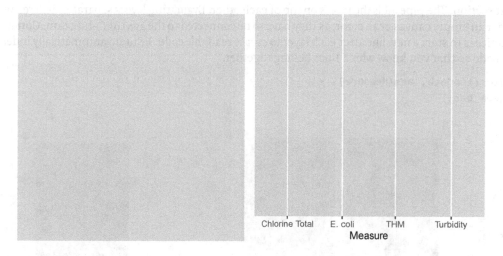

FIGURE 5.4 Empty ggplot canvases.

5.2.2 Aesthetics Layer

The next part in the ggplot() function defines the aesthetics, consisting of the fields in the data used in the visualisation. The example below plots the gormsey data frame and uses the Measure variable (right image in Figure 5.4).

When we add the aesthetics, the ggplot() function draws a canvas with a spot for each of the measures on the *x*-axis. Besides the variables on the axes, you can also define colours, line types, and other data-related design elements in the aesthetics. Still, as discussed in the next section, they will only be visible once you define a geometry. Using `aes(x = Date, y = Result)` will add the Date variable to the *x*-axis and Results to the *y*-axis.

```
ggplot(labdata, aes(Measure))
```

```
ggplot(labdata, aes(Date, Result))
```

5.2.3   Geometries Layer

To visualise the aesthetics, we need to add a geometry that defines the shapes on the canvas. In *ggplot2*, all geometries start with `geom_`. You can explore them easily by typing the first few letters and hitting the TAB button to view the completion options. The *ggplot2* package includes the most common shapes to visualise data, and other developers have shared more specialised geometries. The *ggplot2* website provides detailed descriptions of each available geometry. This chapter introduces some of the most commonly used geometries.

Given that the measure name is a qualitative variable, we can only visualise this data to count the number of samples. Therefore, the `geom_bar()` geometry calculates the number of occurrences of each data point and plots them as a bar chart (Figure 5.5).

The `ggplot()` function passes the first variable in the aesthetics to the geometry. The bar geometry then counts the number of elements for each category in the `Measure` variable. If you don't write anything between the parentheses in the geometry, then the function creates a simple grey chart as in Figure 5.5. The paradigm of maximising the data-pixel ratio suggests that colour should only be used to visualise data. Since we did not instruct R to use colour, it uses grey.

In *ggplot2*, the layers are connected with a plus sign. In computer science terminology, the plus sign is a pipe, which means that the result from the code before the symbol is moved (piped) to the next section. The pipe adds the layers on top of each other. Evaluating layers separately will result in either an empty canvas or an error, as they have to be connected to the `ggplot()` function. Common practice is to start a new line after each layer to create readable code. RStudio automatically indents the code so that you know which lines belong together.

```
ggplot(labdata, aes(Measure)) +
  geom_bar()
```

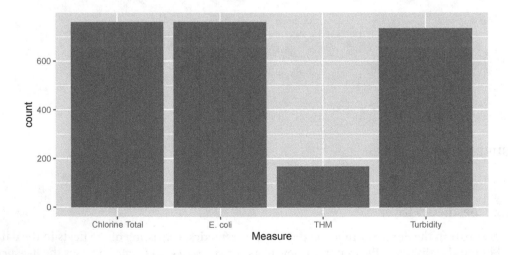

FIGURE 5.5 Bar chart of the number of samples for each suburb.

The *ggplot2* library has a lot of geometries to visualise data, which are used throughout this book. Figure 5.2 shows some of the available geometries.

5.2.4 Colour Aesthetics

Colouring geometries by the data is an extension of the aesthetics layer. Colours are in this case not used for aesthetic reasons, as the name might suggests, but they communicate data.

You can add colour to express data to comply with a style guide by using geom_bar(fill = "royalblue") or any other colour you fancy. Two colour aesthetics are available: col and fill. The former defines the colour of lines, like in Figure 5.6, while the latter describes the colours of shapes. The primary colours, such as red, green, blue, and many more subtle shades, can be called by name. You can see a complete list of the 657 colour names with colors(). The names of colours are, by the way, a fascinating topic in linguistics. The English language has only eleven words that are only used as a colour name, such as red, green, purple. Other colour names are either combinations of words or things that describe a colour, such as dark green, opal white, violet, or turquoise. The number of basic colour terms in other languages is different, showing how the experience of colour varies among cultures (Kay, 2009).

You can also assign a colour with numbers. Various systems are available, with HTML hexadecimal codes as the most commonly used system. HTML hex codes, also called RGB (Red-Green-Blue) triplets, consist of three hexadecimal numbers (ranging from 00 (0) to FF (255)) for each red, green, and blue. The hex code #FF0000 results in a pure red colour, #00FF00 is green. Mixing these three primary colours results in a palette of sixteen million theoretical colours; for example, #78417A forms a deep purple. Adding colour directly into the geometry will give all elements the same colour without expressing data. This method is useful when you comply with a style guide from your organisation or publisher.

Using the aesthetics function, you can assign colours to express data. The example below creates a time series chart of the turbidity results in Gormsey. The col = Suburb part in the aesthetics instructs the ggplot() function to draw a line for each suburb and give it a different colour (Figure 5.6). But first we need to call the *dplyr* library or prepend the function with the package name and two colons to use the filter function. Without this package reference, R will use the filter function in the base package, which will lead to unexpected results.

```
turbidity <- dplyr::filter(labdata, Measure == "Turbidity")
```

```
ggplot(turbidity, aes(Date, Result, col = Suburb)) +
  geom_line()
```

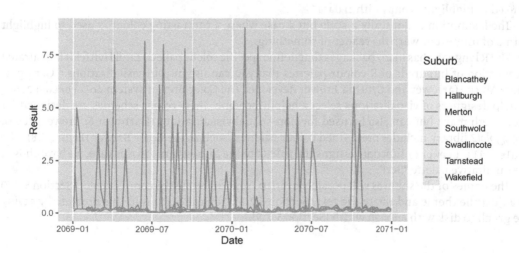

FIGURE 5.6 Gormsey turbidity time series.

This example is not an optimal use of this functionality because there are too many lines, which are hard to read. Referring back to the visual arts, this graph is a bit like a Jackson Pollock action painting. We can fix this with the next layer, the facets (Section 5.2.5).

The *ggplot2* package assigns a default scheme to each colour aesthetic, which you can obviously change. Choosing the optimal colour scheme is part art and part science. There are basically four types of colour schemes, shown in Figure 5.7.

Sequential schemes contain a series of colours with increasing strength, ranging from pale to the darkest blue. These colour schemes are most suitable for visualising low to high magnitude, with light colours for low data values and dark colours for high values. The dark colour will grab the viewer's attention, while the light colours fade into the background.

This scheme is often used in business reports. Diverging colours visualise deviations from a norm, such as droughts, floods, or water consumption, using two sequential colours and a neutral midpoint. Green, amber, and red are the most common use of this type of palette. This type of reporting highlights anomalies and problem areas to discuss actions to improve future performance. A note of caution is that this technique does not work for people with green/red colour blindness. This condition is not a problem with real traffic lights as the order of the lights is always the same. However, on a business report, the colours will all look the same to roughly eight percent of men and one percent of women with this condition (Wexler et al., 2017).

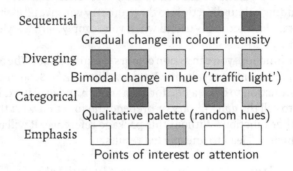

FIGURE 5.7 Data visualisation colour scales.

Qualitative colours are aesthetically compatible palettes without a logical relationship with the data. These palettes can express qualitative values such as categories, such as the name of a suburb or used to highlight groups within data.

The last option is not really a scale but a case where a contrasting colour is used to highlight points of interest or warn the reader of something.

The R language has many packages that define specific colour pallets. Emil Hvitfeldt has curated a collection of hundreds of R colour palettes that you can use in your visualisations.[2] Cartographers Mark Harrower and Cynthia Brewer developed the Color Brewer system colorbrewer2.org to help designers of visualisations select a helpful scheme. These colour schemes are designed for choropleth maps but can also be used for non-spatial visualisations (Harrower & Brewer, 2003). The *ggplot2* library includes predefined colour palettes, such as the Color Brewer (Figure 5.8). The scale_fill_brewer() function assigns the palette to the fill aesthetic. In this case, we have chosen the qualitative palette "Set1".

The names of the suburbs will possibly overlap. You can rotate the complete plot (Section 5.2.7) or adjust the theme and rotate the text (Section 5.2.8). You can also avoid this problem by saving the graph to disk with enough width (Section 5.2.9).

---

[2]Emil Hvitfeldt, R Palette Picker (emilhvitfeldt.github.io/r-color-palettes/).

```
ggplot(labdata, aes(Suburb, fill = Measure)) +
  geom_bar() +
  scale_fill_brewer(type = "qual", palette = "Set1")
```

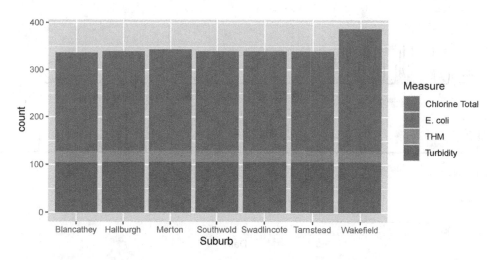

FIGURE 5.8 Color Brewer example.

You can see all the palettes in the Color Brewer schemes by invoking the *RColorBrewer* package (Neuwirth, 2022):

```
RColorBrewer::display.brewer.all()
```

Various palettes are available, which you can find using the completion system. Note that when you use the col aesthetic for lines, you need to use the scale_color_brewer() function to assign a palette. In some visualisations, you might define a different palette for fills and lines by using both versions.

The scale_fill_manual() and scale_color_manual() functions let you define your own scheme, for example the corporate style guide. You define the colour with the values parameter. You just need to ensure that the number of colours matches the number of unique values in your aesthetics. You can use the colour names or their hexadecimal values mentioned above, for example:

```
ggplot(labdata, aes(Suburb, fill = Measure)) +
  geom_bar() +
  scale_fill_manual(values = c("cornflowerblue",
                               "darkseagreen",
                               "#ee6611",
                               "#ccaa44"))
```

5.2.5 Facets Layer

The time series plotted in Figure 5.6 was confusing because there were too many lines on the canvas. A facet allows us to combine multiple plots in one visualisation. Facets are an efficient and commonly used method to communicate multivariate data within one visualisation without compromising readability. The example below uses the facet_wrap() function to create separate time series for each suburb. The wrap function finds the best way to fit them on the canvas (Figure 5.9). Note the tilde at the start of the facet parameter.

```
ggplot(turbidity, aes(Date, Result)) +
  geom_line() +
  facet_wrap(~Suburb)
```

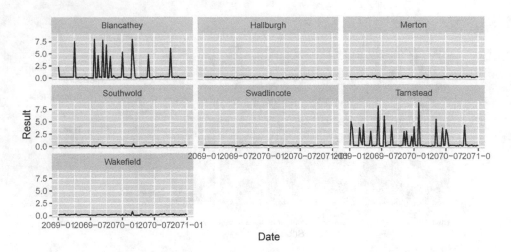

FIGURE 5.9 Faceted plot of turbidity in water quality zones.

The `facet_wrap()` function lists the individual graphs in rows. You can control the number of columns with the `ncol` parameter.

5.2.6   Statistics Layer

Now that we have a powerful tool to visualise data, we also want to add elements to tell a story. At this point, our time series is just a series of lines without context. The statistics layer adds summaries and reference points for the viewer to see the data in context.

The example in Figure 5.10 adds two further layers to add context. The first three lines of the code create a data frame with the highest measured THM value for each day. Next, the `geom_smooth()` function draws a regression line on the canvas. The Loess function is the default method, but it can also be used for linear and regression models with the `method` parameter. Note how the smoothing line is drawn first, so it stays in the background. Subsequent geometries are plotted on top of the previous ones. The story is completed by drawing a red horizontal line at 0.25 mg/l, the regulatory limit, using the `geom_hline()` geometry. You can add vertical lines with the `geom_vline()` geometry. We have now reached a point where a graph tells a complete story. This visualisation tells us that there is a flat trend in THMs. Still, we had several spikes above the regulatory limits.

```
library(dplyr)
thm <- filter(labdata, Measure == "THM")
thm_grouped <- group_by(thm, Date)
thm_max <- summarise(thm_grouped, thm_max = max(Result))

ggplot(thm_max, aes(Date, thm_max)) +
  geom_smooth(method = "lm") +
  geom_line() +
  geom_hline(yintercept = 0.25, col = "red")
```

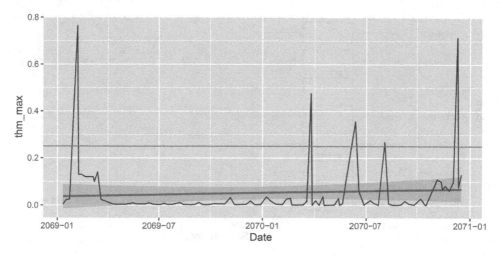

FIGURE 5.10 THM trends.

5.2.7 Coordinates Layer

The penultimate layer defines how we display the coordinate system. The *ggplot2* package can show the standard Cartesian coordinates, polar coordinates, and various mapping projections. For now, we will stick to the Cartesian system. All mapping layers star with coord_, which you can explore the usual way with the completion system.

Section 4.7.2 discussed box plots, which you can also create with *ggplot2* (Figure 5.11). If we pass one value to the box plot aesthetic, only one box is plotted. Adding a second value groups the data by that variable and plots multiple boxes. The third line introduces a new layer that converts the *y*-axis to a logarithmic scale because the data is positively skewed. Without this transformation, the visualisation will be challenging to read.

```
ggplot(turbidity, aes(Suburb, Result)) +
  geom_boxplot() +
  scale_y_log10(name = "Samples (log)", n.breaks = 10) +
  coord_flip()
```

You can define the *x*- and *y*-axes with the scale_* set of functions. This functionality transforms scales, but it also allows you to define how the scale is displayed. Some of the most common functions in this group are:

- scale_x_log10(): Logarithmic scale

- scale_x_discrete(): Discrete variables (names)

- scale_x_continuous(): Continuous variables, such as measurements

- scale_x_date(): For displaying dates and times

The scale layer also sets the name for the *y*-axis to "Samples" and displays up to ten breaks. Further parameters are available to fine-tune how axes are displayed. More advanced examples in the following chapters provide some more guidance. The coord_flip() transformation rotates the graph. This layer was needed because the names of the suburbs can overlap when the *x*-axis is horizontal.

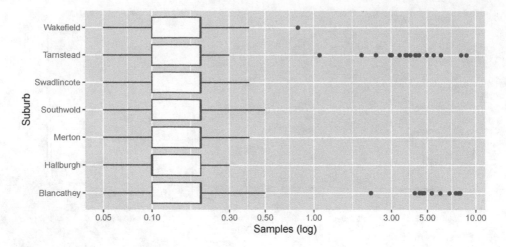

FIGURE 5.11 Distribution of turbidity results.

To prevent the overlap of dates in the time-series graph in Figures 5.9 and 6.4, we need to define how dates are displayed. Adding the `scale_x_date(date_breaks = "1 years", date_labels = "%Y")` layer to the plot will only show a tick every year, and the label is reduced to just the four digits of the year (`"%Y"`). R has extensive capabilities to format dates, which are explained in more detail in Chapter 11.

5.2.8 Theme Layer

The theme of a graph defines the aesthetic aspects of the background, fonts, axes, and so on. To use one of the predefined themes, simply add `theme_name()` to the function call and replace `name` with the name of the theme (Figure 5.12).

```
ggplot(turbidity, aes(Date, Result)) +
  geom_line() +
  facet_wrap(~Suburb, ncol = 1) +
  theme_void(base_size = 12)
```

You can try different themes by typing `theme_` and hitting TAB to see the available themes. However, the text in a plot can sometimes be a bit small, and you can change this with the `base_size = x` option, where `x` is an integer.

The *ggplot2* library has extensive options to change the theme of a graph. Every canvas element can be changed, colours and lines, text sizes, fonts, and so on. Modifying themes is a complex topic due to the countless variations.

The example below rotates the text on the *x*-axis by ninety degrees. The theming capabilities in the *ggplot2* package allow you to create corporate templates to conform with the relevant style guide. The *ggplot2* documentation provides detailed information about how to change every aspect of your theme and how to save and set new defaults.

```
ggplot(labdata, aes(Measure)) +
  geom_bar() +
  theme(axis.text.x = element_text(angle = 90))
```

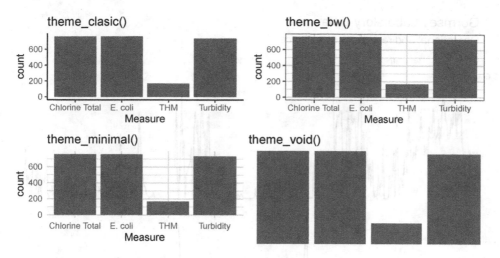

FIGURE 5.12 Some examples of ggplot themes.

5.2.9   Sharing Visualisations

The last step is to tell the data story and share your visualisation. The graphs we have created don't include much context to know what we are looking at. The labs() function helps add text to the plot and change the axes' labels, as shown in the example below (Figure 5.13). Adding text to the plot prevents confusion if the file is separated from its context. As a final example, this graph adds almost all elements discussed in this chapter and summarises how the laboratory data compares to the regulatory limits. To save some space and prevent cramped graphics, this version only shows three cities and omits the E. coli measure. The %in% function is a convenient way to filter by a list of possible options. The limits data frame contains the applicable limits and adds this as a separate layer. Each geometry can have its own data layer and aesthetics using the data and aes parameters.

```
limits <- data.frame(Measure = c("Chlorine Total", "Chlorine Total",
                                 "THM", "Turbidity"),
                     Limit = c(1, 2, 0.25, 5))

ggplot(filter(labdata, Measure != "E. coli" &
                       Suburb %in% c("Merton", "Tarnstead", "Blancathey")),
       aes(Date, Result)) +
  geom_hline(data = limits, aes(yintercept = Limit),
             col = "red", linetype = 2) +
  geom_line(size = .5) +
  facet_grid(Measure ~ Suburb, scales = "free_y") +
  scale_x_date(date_labels = "%Y", date_breaks = "2 years") +
  theme_minimal(base_size = 11) +
  labs(title = "Gormsey Laboratory Data",
       subtitle = "Operational and regulatory compliance",
       caption = "Source: Gormsey Laboratory")
```

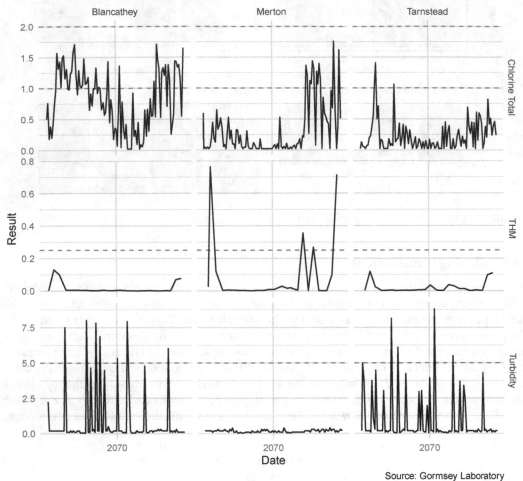

FIGURE 5.13 Visualising all laboratory data results in a grid.

Anyone looking at the graph in Figure 5.13 has all information available to draw conclusions about the performance of these systems. The chart is easy to read, and the criterion for what constitutes a spike is also visible.

Showing the graphs on the screen is fine, but you will most likely want to share them with colleagues. The ggsave() function provides a convenient method to save a *ggplot2* graph to a file in PNG, PDF, JPG, or many other formats. The ggsave() function always saves the most recent plot. The default settings save the figure at a high resolution of 300 Dots Per Inch (DPI), suitable for printed publications. The width and height, unfortunately, default to inches. You can also set the units parameter to change this to "cm", "mm", or pixels ("px"). The dpi option sets the pixel density in dots per inch for the saved plot if saved as a raster image. A density of 300 dpi or greater is print quality. When preparing visualisations for printed publications, best practice is to save them in postscript (.ps) or PDF.

```
ggsave("gormsey_labdata.pdf", width = 297, height = 210, units = "mm")
```

5.3   **Further Study**

This chapter only provides a brief overview of *ggplot2*, highlighting the principles of visualisation. However, the capabilities of this package are extensive and far beyond what this one chapter can cover. Further examples in this book show more use cases for this package. In addition, the Tidyverse website contains comprehensive information about the *ggplot2* package. The R Charts website (r-charts.com) provides a comprehensive overview of the visualisation types in base R, *ggplot2*, and other packages.

The visualisation tools in base R are often great tools for quick review of the data. The *ggplot2* tools are more suitable for presentations. It is interesting to note that *ggplot2* does include any geometry for a pie chart (Table 5.1). Pie charts are usually discouraged as a useful visualisation tool, especially when there are more than four slices (Wong, 2010).

TABLE 5.1 Comparison between base R and ggplot2 visualisations.

Type	Base R function	*ggplot2*
Scatter plot	`plot(x, type = "p")`	`geom_point()`
Line plot	`plot(x, type = "l")`	`geom_line()`
Histogram	`hist(x)`	`geom_histogram()`
Box plot	`boxplotx)`	`geom_boxplot()`
Bar plot	`barplot(x)`	`geom_bar()`
Pie plot	`pie(x)`	Not available

Several packages exist that extend the capabilities of the *ggplot2* package by providing additional geometries and themes, some of which are used in the remainder of this book. The names of these packages generically start with *gg* (Grammar of Graphics), so they are easy to find. The advantage of visualisation tools that follow the *ggplot2* principles is that you can combine the various geometries and packages with telling your story.

The plots we have seen are static and don't allow the user to explore the data. However, several packages are available, such as *plotly*, to create graphics that the user can explore (Sievert et al., 2022). Most of these visualisations are HTML files you can view in your internet browser.

Sharing a PDF file with visualisation is fine, but it still means that you need to manually insert these in your final report. The next chapter discusses a reproducible method to share the results of your analysis with colleagues or the general public with a PowerPoint slide deck or Word document.

6

Sharing Results

Data science aims to create value from data by creating useful, sound, and aesthetic data products. Analysing data is rewarding, but creating value requires you to communicate the results. Analysing data in RStudio is fun but hard to share with anyone who does not understand the language. Of course, you could copy and paste results into a document, but that is not very efficient or reproducible. This chapter explains how to communicate the fruits of your labour with colleagues and other interested parties by generating reports that combine text and analysis through literate programming.

The generic term for the result of a data science project is a data product, which can be a presentation, a written report, article or book, an infographic, a computer application or anything else that communicates the analysis results. Chapter 1 mentioned how data products need to be reproducible and replicable. This means that reviewers can reproduce the process by following the code and that it can be easily replicated when new data becomes available. Reproducibility is an essential dimension for quality assurance and academic peer reviews. Replicability creates business efficiency by reusing a data product for new situations. The learning objectives for this chapter are:

- Implement the workflow for data science projects

- Apply the principles of reproducible and replicable research

- Use basic R Markdown to create a PowerPoint presentation from data

6.1 Data Science Workflow

A successful data science project follows a defined workflow from problem definition to communicating the results (Wickham, 2016), visualised in Figure 6.1. A well-defined problem will make the rest of the process much easier and prevents data dredging. The next step involves loading and transforming the data into a suitable format for the required analysis. The core of the data science workflow is an iterative loop, the so-called data vortex. This vortex consists of three steps: exploration, modelling, and reflection. The analyst repeats these steps until the problem is solved or found to be unsolvable, in which case the problem needs to be stated differently. The final stage of the workflow involves communicating the results.

6.2 Define

The first step of the workflow describes the problem under consideration and the desired future state. Analysing data is not about the numbers but about creating actionable insights, as illustrated

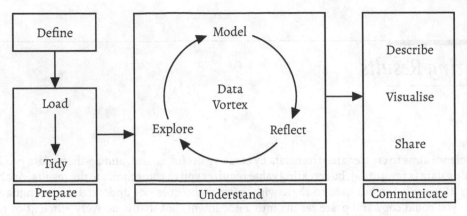

FIGURE 6.1 Data science workflow.

in Figure 1.2. Data analysis is practical when it either contains an actionable insight that improves reality or provides assurance that a process operates within expected parameters.

The problem definition should not refer to available data or possible methods but focus on the issue. Selecting the right data and techniques is secondary to choosing the right questions (Shron, 2014). For example, an organisation could seek to optimise production facilities, reduce energy consumption, monitor effectiveness, understand customers, and so on. In other words, the problem definition needs to be rooted in reality.

A concise problem definition is necessary to ensure that a project does not deviate from its original purpose and consists of four parts:

1. Context: What is the background of the problem?
2. Needs: What are the pain points?
3. Vision: What does success look like?
4. Outcome: What is the data product?

Context

To ensure water safety, most water utilities add chlorine to remove pathogens. However, this addition has the unintended consequence of decreasing the taste experienced by customers. Although chlorination is often necessary to ensure safe water, it is a counter-intuitive promise of safety for customers and does not relate to good water in public perception (Parr, 2005; Prevos, 2017).

Needs

The acceptability of tap water decreases as the level of chlorine increases. The threshold for chlorine detection is about 0.25 mg/l Cl_2. At about 0.6 mg/l, less than half of the consumers found the taste acceptable (Puget et al., 2010). The World Health Organization recommends that total chlorine levels are between 0.2 and 1.0 mg/l at all times to ensure tap water is safe to drink.

Vision

The Gormsey water utility has received numerous complaints about water taste in some suburbs and needs to better understand the relationship between customer perception and chlorine levels.

Outcome

Your task is to explore laboratory data and estimate the likelihood that somebody has a negative taste experience when drinking tap water.

The remainder of this chapter will use this problem statement as a case study, ending with a script that creates a PowerPoint presentation of the analysis. The analysis in this chapter is not an exhaustive solution to the tensions between the customer experience and public health concerning chlorine dosing. Instead, the analysis in this chapter is an example of how to implement the steps in the workflow and how to use literate programming to combine code and text.

6.3 Prepare

The available data needs to be loaded and wrangled into the required format before any analysis can occur. Anecdotally, this project phase could consume up to eighty per cent of the work effort. The data from the first case study requires almost no transformation, but that is rarely the reality. Chapter 7 discusses some basic techniques to clean data.

Best practice in data science is to describe every field used in the analysis to ensure the context in which the data was created is understood. Defining a transparent methodology is where the science comes into data science and is the point where it distinguishes itself from traditional business analysis.

We have the Gormsey data explored in the previous chapters for our case study. This data is already tidy and does not need any cleaning, so we can proceed without further ado. Tidy data is a standard way of mapping the meaning of data to its structure. Chapter 7 further explains the concept of transforming data to an ideal state. All we have to do for this case study is filter the data to contain only the chlorine values.

The data and code for this session are available in the RStudio project. Section 2.3.1 describes how to obtain these files.

```
library(tidyverse)
labdata <- read_csv("data/water_quality.csv")
chlorine <- filter(labdata, Measure == "Chlorine Total")
```

6.4 Understand

Once the data is available in a tidy format, understanding the data can commence. The analytical phase consists of the three-stage data vortex. These three stages are: explore, model, and reflect. The term vortex is descriptive because the reality of analysing data is rarely as linear as the diagram suggests. Depending on the complexity of the data and the problem, this part of your journey will involve a lot of trial and error.

The techniques used in this phase depend on the type of problem that is being analysed. Also, each field of endeavour uses different methodological suppositions and methods. For example, analysing subjective customer data requires a different approach than the objective reality of laboratory results. For our simple case study, the analysis only consists of descriptive statistics.

6.4.1 Explore

The first step when analysing data is to understand the relationship between the data and the reality it describes. Generating descriptive statistics such as averages, ranges, distribution curves, and other summaries provides a quick insight into the data (Chapter 4). However, relying on numerical analysis alone can be deceiving because ostensibly different data sets can result in the same values, as we saw with Anscombe's quartet (Figure 4.6 on page 46).

Justin Matejka and George Fitzmaurice have taken this idea to the next level (Matejka & Fitzmaurice, 2017). Figure 6.2 shows four of twelve data sets with different geometric patterns. However, the descriptive statistics of *x* and *y*, and their correlations are almost identical for all subsets. You can explore the data with the *datasauRus* package (Davies et al., 2022).

FIGURE 6.2 Four scatter plots with similar summary statistics.

Another reason visualisations are essential to explore data is to reveal anomalies, such as spikes in time series or outliers. A sudden increase and decrease in physical measurements is often caused by issues with probes or data transmission instead of actual changes in reality. These spikes need to be removed to ensure the analysis is reliable. Anomalies in social data, such as surveys, could be from subjects that provide the same question to all answers or missing answers.

Detecting and removing outliers and anomalies from the data increases the reliability of the analysis (refer to Chapter 12). However, not all oddities in are necessarily suspicious, and care should be taken before removing data. Furthermore, the reasons for excluding any anomalous data should be documented so that the analysis remains reproducible and can be corrected if later found to be wrong. Chapter 7 discusses dealing with missing data in further detail.

For our case study, we could use a boxplot for each suburb to provide a concise summary of the statistical distribution of chlorine results and indicate which values are outside aesthetic or health limits. The standard `boxplot()` function will suffice as we only explore the data. For publication-ready plots, the *ggplot2* package might be more helpful (Figure 6.3).

```
par(mar = c(8, 4, 4, 1))
boxplot(Result ~ Suburb, data = chlorine, pch = 19, cex = .5,
        las = 2, xlab = NULL, ylab = "Chlorine level [mg/l]")
abline(h = c(0.25, 0.6), col = "red", lty = 2)
```

This visualisation shows that the suburbs of Blancathey and Wakefield present significantly higher chlorine levels compared to the other suburbs. The horizontal lines indicate the taste threshold level and the level at which chlorine can make the water less pleasant to drink. The data shows that in most cases, the chlorine will be noticeable when drinking the water. But this graph does not tell us much on how often a very high level of chlorine occurs.

The faceted graph in Figure 6.4 provides some insights into chlorine levels over time. The code below creates a time series of the chlorine levels for all suburbs in Gormsey with the critical upper limit for the taste experience. The `scale_x_date()` layer converts the date variable to just the year, which is further explained in Chapter 11.

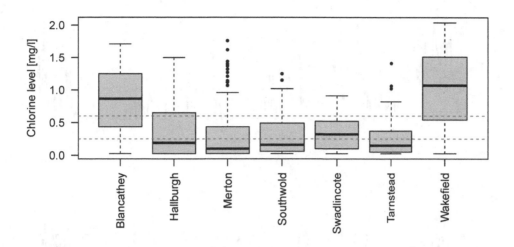

FIGURE 6.3 Chlorine sample results distribution.

```
ggplot(chlorine, aes(Date, Result)) +
  geom_line() +
  geom_hline(yintercept = 0.6, col = "red") +
  scale_x_date(date_breaks = "year", date_labels = "%Y") +
  facet_wrap(~Suburb) +
  theme_minimal() +
  labs(title = "Gormsey chlorine levels", x = NULL)
```

This exploration shows that all suburbs have levels of chlorine that could cause a negative taste experience, with Wakefield as the outlier, which could mean it is closer to the treatment pant than the other suburbs. The exploration also shows that Merton and Tarnstead have issues with low levels of chlorine. Now that we explored the data distribution, let's move to some modelling.

6.4.2 Model

After the analyst has obtained a good grasp of the variables under consideration, the actual analysis can commence. Modelling involves transforming the problem statement into mathematics and code. Every model is bounded by the assumptions contained within it. Statistician George Box is famous for his idea that "Models, of course, are never true, but fortunately, it is only necessary that they be useful" (Box, 1979). Data science is not a science in that we seek some universal truth by building perfect models of the world. We can only hope to find something useful that can positively influence reality.

The original research question must always be kept in mind when modelling the data. Exploring and analysing data without a specific purpose can quickly lead to wrong conclusions. Just because two variables correlate does not imply a logical relationship. A clearly defined problem statement and methodology prevent data dredging.

The abundance of data and the ease of extracting information makes it easy for anyone to find relationships between different sources of information. The *Spurious Correlations* website hosts a delightful collection of strong but nonsensical correlations. Did you know that the per-capita cheese

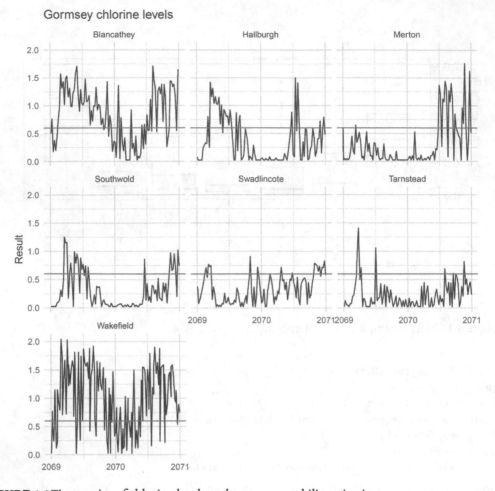

FIGURE 6.4 Time series of chlorine levels and taste acceptability criterion.

consumption in the US correlates strongly with the number of people who die by becoming tangled in their bedsheets?

This problem is what Drew Conway coined the danger zone in his data science Venn diagram (Figure 1.1). While it is easy to visualise data, we should never draw conclusions without understanding the mathematical principles at work. A good general rule when analysing data is to distrust your method when you can easily confirm your hypothesis.

This simple case study does not require advanced modelling, as the problem statement is descriptive. This code summarises the chlorine levels by suburb and calculates the percentage of samples above the taste-acceptance criterion of 0.6 mg/l. The arrange() function sorts the output from low to high.

```
chlorine_suburb <- group_by(chlorine, Suburb)

chlorine_results <- summarise(chlorine_suburb,
                        Minimum = min(Result),
                        Mean = mean(Result),
                        Max = max(Result),
                        Taste = round(sum(Result > 0.6) / n() * 100))
```

```
arrange(chlorine_results, Taste)
```

```
# A tibble: 7 × 5
  Suburb      Minimum  Mean   Max Taste
  <chr>         <dbl> <dbl> <dbl> <dbl>
1 Tarnstead     0.025 0.225  1.41     7
2 Merton        0.025 0.311  1.76    16
3 Swadlincote   0.025 0.339  0.91    16
4 Southwold     0.025 0.308  1.25    22
5 Hallburgh     0.025 0.373  1.5     26
6 Blancathey    0.025 0.824  1.71    63
7 Wakefield     0.025 1.00   2.04    71
```

6.4.3 Reflect

Before you can communicate the results of an analysis, domain experts need to review the outcomes to ensure they make sense and solve the problem stated in the definition. The reflection phase should, where relevant, also include customers to ensure that the problem statement is being resolved to their satisfaction.

Visualisation is a quick method to establish whether the outcomes make sense by revealing apparent patterns. Another powerful technique to reflect on the results is sensitivity analysis. This technique involves changing assumptions to test that the model responds as expected.

Reflecting on the analysis of this case study does not require much complexity. Of course, the analysis could be more sophisticated, but this chapter is about the principle, not the problem itself.

6.5 Communicate

The last, arguably, most challenging phase of a data science project is communicating the results to the users. "The data says" is an often-heard claim that is incorrect. Data cannot speak for itself; only people can say things about this word (Anderson, 2015). This statement might be a cliche, but the basic principle is valuable. Communicating results is an essential skill for anyone analysing data.

In most cases, the users of a data product are not specialists with a deep understanding of data and mathematics. The skill difference between the data scientist and the user of their products requires careful communication of the results.

To claim that a report needs to be written with clarity and correct spelling and grammar almost seems redundant. The importance of readable reports implies that the essential language a data scientist needs to master is not Python or R but English, or whatever language you communicate in. Writing a good data report enhances the reproducibility of the process by describing all the steps. A report should also help to explain any complex analysis to the user to engender trust in the results.

The topic of writing useful business reports is too broad to do justice within the narrow scope of this book. However, for those people that need help with their writing, data science can also assist. Many great online writing tools support authors with spelling, grammar, and idiom. These advanced spelling and grammar checkers use advanced text analysis tools to detect more than spelling mistakes and can help fine-tune a text utilising data science. However, even grammar checking with machine learning is not a perfect replacement for a human being who understands the meaning of the text.

6.5.1   Reproducible and Replicable Research

Good data science is reproducible and replicable. This means that the process of raw data to draw conclusions is transparent so that the data, the code, and the text are available. Research is reproducible when you repeat the same analysis with the same data. Reproducibility allows for more comprehensive peer reviews. Research is replicable when you can perform the same analysis with different data. Replicability enhances the efficiency and reliability of your analysis as you only need to manage one source of computation. These definitions are in flux within the data science community. The definitions in this book (Figure 6.5) follow the approach used in the *Turing Way* project (The Turing Way Community, 2021).

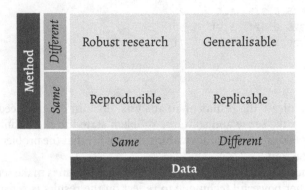

FIGURE 6.5 Types of reproducibility.

Chapter 5 showed how to use the *ggplot2* package to create aesthetic visualisations and save them to disk in a high resolution with the ggsave() function. You can then load these images in your report to communicate the results. This approach works fine until you need to change some assumptions in your graphs, a new colour scheme, or any other change. Every time you change the analysis, you need to edit the report. This inefficient workflow leads to errors as you might forget to transpose the new results into the report. The problem with the traditional approach is that the data and the code are separated from the final data product. Reproducible research solves this problem by combining the data and the analysis with the final result.

The most effective method to achieve full reproducibility is to use literate programming. Although graphical programs might at first instance appear more user-friendly than writing code, point-and-click systems have severe limitations. Most importantly, the results are often impossible to verify as there is no sequential script that shows the steps of the analysis. The central concept of literate programming is that the data, the code, and the output are logically linked so the result will also change when the data or the code changes.

The primary method to make your code reproducible is adding comments to the code. Comments help a human reader understand the flow of logic. There is a point, however, where your analysis is so complex that you need more comments than code. When this is the case, you need to use more advanced methods. Furthermore, most consumers of data products are not interested in the code and only want to see the results.

6.5.2   Workflow for R Coding

The workflow from raw data to a final data product, such as an article, book, or presentation, requires three coding steps (Figure 6.5) from the raw data to the end product.

The processing code transforms the raw data into a suitable format for analysis. Chapter 7 shows how to develop such a script. Next the analytic code creates the outputs from the cleaned data. The

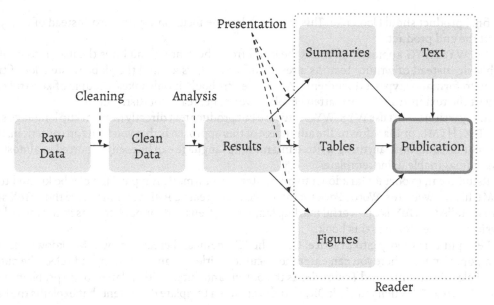

FIGURE 6.6 Reproducible research pipeline.

presentation code converts these outputs to figures, tables, and numerical summaries (Chapter 5). This code is merged with the text in the final step to create the data product.

In traditional publications, the reader only sees the results of the analysis expressed in the text, tables, and figures. The author is the only one who understands the steps from raw data to the final product. Best practice in academic writing is that authors provide the raw data, the code, and the finished article so that an interested reader or peer reviewer can reproduce the complete analysis. Providing data and code helps readers and reviewers to understand the text beyond their presented results and promotes a high level of peer-review (Peng, 2011).

Several methods exist to write presentation code to convert data, text, and code into a publication. The most popular method for the R language is R Markdown, which allows you to combine text, images, and code to create presentations, articles, websites, or books. Literate programming is the most efficient method of writing.

6.6 R Markdown

R Markdown is a method to combine a narrative with the results of the analysis. An R Markdown (Rmd) file consists, as the name suggests, of R code and Markdown code. You already know what R is, so now we need to explore Markdown (Xie, 2018).

Contemporary software follows the *What You See Is What You Get* (WYSIWYG) principle. Graphical interfaces simulate the physical world by making objects on the screen look like pieces of paper and folders on a desk. You point, click, and drag documents into folders; documents appear as they would on paper and when done, they go into the rubbish bin. Graphical interfaces are magic tricks that make you believe you are doing physical things. Unfortunately, this approach moves people away from understanding how a computer works.

RStudio and other text editors use the *What You See Is What You Mean* (WYSIWYM) principle. As I write this book, I don't see what it will look like in printed form as you would using modern word processors. In RStudio, I only see text, images, and some instructions for the computer on what the final product should look like. This approach lets me focus on writing text instead of worrying about the end product.

The WYSIWYG approach distracts the mind from the content and lures the user into fiddling with style instead of writing text. As a result, office workers around the globe waste a lot of time trying to format or typeset documents. As I write this book, it only takes a couple of keystrokes to convert the text into a fully formatted ebook or web page, ready for distribution.

Many writers don't use WYSIWYG software but produce text directly in a markup language such as LaTeX, HTML, or Markdown. The advantage of this approach is that you focus on content instead of design. In addition, anything written in a markup language can be easily exported to almost any format imaginable using templates.

Rstudio can export R Markdown to many standard formats. A typical file can be knitted to an HTML file for websites, Word, PowerPoint, or PDF. To create a PDF, you must have the LaTeX software installed. LaTeX is a powerful markup language, often used for publications in the formal and physical sciences, such as this book.

Let's put this theory into practice. Go to the *File* menu and create a new R Markdown file. You see a popup menu where you can enter the document title and author name, and select the output type. Select *Presentation* and *PowerPoint* as the output and enter a title related to the problem statement (Figure 6.7). When you click OK, RStudio creates a template document that explains the basic principles.

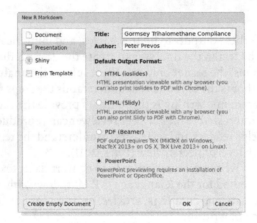

FIGURE 6.7 R Markdown popup menu (Source: cloud.rstudio).

When you click the *Knit* button, RStudio asks you to save the file and generates a document that includes the written content, some of the code, and the output of any R code embedded in the report. When you do this for the first time, you might receive a message that specific packages need to be installed. Follow the prompts to let that happen. An R Markdown document consists of three elements:

- Metadata

- Formatted text

- Code (chunks)

6.6.1   Defining Metadata

The first few lines of the new document are the metadata. This is where you define the title, author name, date, and output format. This data is copied from the popup menu, but you can edit it between the three dashes. This lines are YAML, which stands for *AML Ain't Markup Language*, a common method to define variables in plain text files.

The header defines the title and author for the document. The date uses an R function that displays the current system date. Finally, we define the output as a PowerPoint presentation and select a template to control the design of the slides. Note the indentation of two and four spaces at the start of the last two lines. You need to follow this precisely for RStudio to recognise the data structure.

```
---
title: "Negative Taste Experiences in Gormsey"
author: "Peter Prevos"
date: "`r Sys.Date()`"
output:
  powerpoint_presentation:
    reference_doc: ../data/template.potx
---
```

The following line sets the overall parameters for the document. These parameters determine how the R code is evaluated and presented. Any R code must be embedded in a 'chunk' marked by three backticks and settings between curly braces. You can find this symbol under your escape key.

This line of R code defines the defaults for all following chunks. In this example, echo = FALSE means that the output document does not include the code. This code also silences any warnings and messages R usually shows in the console. The dpi argument ensures high resolution graphics. The include option excludes the code itself from the presentation.

Many other configuration options are available, which are explained in the book *R Markdown: The Definitive Guide* by Xie (2018). The book is freely available on the internet on the bookdown.org website. This website contains many books written with R Markdown.

```
```{r setup, include=FALSE}
knitr::opts_chunk$set(echo = FALSE, warning = FALSE, message = FALSE, dpi = 300)
```
```

6.6.2 Formatted Text

The third part of the document is text in Markdown format, which looks something like this:

```
# Problem Statement
- Chlorine levels about 0.6 mg/L are acceptable by only 50% of customers
- A minimum chlorine residual is required for public health
- Exploratory analysis of chlorine levels for all Gormsey suburbs
```

When knitting this text, every line that starts with a hashtag (#) will become a new slide. In text documents it becomes a header. The lines starting with a dash will become dotpoints. Several symbols are available to control how text is presented. The list below shows some of the most common syntax in Markdown:

- Headings: `* H1, ** H2` etc.

- Bold: `**bold text**`

- Italic: `*italic text*` or `_italic text_`

- Blockquote: `> blockquote`

- Ordered List:

- 1. First item
- 2. Second item
- 3. Third item

- Unordered List:

  - - First item
  - - Second item
  - - Third item

- Code: 'code'

- Horizontal Rule: --

- Link: [title](https://www.example.com)

- Image: ![alt text](image.jpg)

When using PowerPoint as the output format, level 1 (#) or level 2 (##) headings (when this is the highest level) indicate new slides.

6.6.3   Code Chunks

The most essential parts of an R Markdown document are the code chunks that analyse the data and produce the output. The output of these functions will be knitted into the PowerPoint document.

You can add additional chunks with the insert button or by pressing `control-alt i`. When you click the knit button, you notice that RStudio can also process other data science languages such as SQL or Python.

The data folder is preceded by two dots refer to the folder above the working directory. When working with R Markdown documents, the working directory is the folder the file is located in, which requires the two dots to ensure R can find the file. The same method is used in the metadata chunk discussed above.

```
# Laboratory Data
```{r}
library(tidyverse)
labdata <- read_csv("../data/water_quality.csv")
chlorine <- filter(labdata, Measure == "Chlorine Total")
```
```

The rmarkdown package has a lot of options to control how the code is evaluated and the output formatted. You can set them for each chunk or set them as defaults in the first code chunk.

- `echo`: Show or hide the code itself (TRUE or FALSE)

- `fig.width` and `fig.height`: Size of the plots in inches

- `message`: Show or hide messages

- `warning`: Show or hide warnings

The code block below shows how to insert a visualisation onto a slide, matching the dark background of the PowerPoint template.

```
# Explore results
```{r, fig.width=9, fig.height=4}
ggplot(chlorine, aes(Date, Result)) +
    geom_line(col = "white") +
    geom_hline(yintercept = 0.6, col = "red") +
    scale_x_date(date_breaks = "year", date_labels = "%Y") +
    facet_wrap(~Suburb) +
    theme_dark() +
    theme(plot.background = element_rect(fill = "black"),
          title = element_text(colour = "white"),
          axis.text = element_text(colour = "white")) +
    labs(title = "Gormsey chlorine levels", x = NULL)
```
```

R evaluates the expression when you knit the document. This way, your numbers are always up to date with the latest data. Ensure you don't forget the lower case letter r to indicate that it needs to be evaluated.

6.6.4 Formatting Tables

The output of data analysis is often expressed in tables. To create neat tables in a report, you should use the kable() function of the *knitr* package . This example in Table 6.1 shows how to export a nicely-formatted table in an R Markdown file.

```
# Top-three High Chlorine Locations
```{r}
chlorine_suburb <- group_by(chlorine, Suburb)

chlorine_results <- summarise(chlorine_suburb,
                    Minimum = min(Result),
                    Mean = mean(Result),
                    Max = max(Result),
                    Taste = round(sum(Result > 0.6) / n() * 100))

knitr::kable(top_n(arrange(chlorine_results, desc(Taste)), 3), digits = 2,
             col.names = c("Suburb", "Minimum Cl", "Mean Cl",
                           "Maximum Cl", "Negative taste %"))
```
```

TABLE 6.1 Knitr table example.

Suburb	Minimum Cl	Mean Cl	Maximum Cl	Negative taste %
Wakefield	0.03	1	2.04	71
Blancathey	0.03	0.82	1.71	63
Hallburgh	0.03	0.37	1.5	26

The *knitr* library provides functions to knit R and Markdown into the preferred output. The second line filters the Gormsey laboratory data with failed THM tests. Finally, the last line converts this data frame into a well-formatted table for MS Word, PowerPoint, HTML, or PDF.

For more precise control of the design of tables, use the *Flextable* package (Gohel & Skintzos, 2023). This library provides fine-grained control over fonts, colours, sizes, and anything else you like to play with.

6.6.5   Inline Code

Lastly, you can embed the output of an R expression inside a line of text. For example, to write: "A total of 760 chlorine results appear in the data." To achieve this, you can embed R code within a sentence: `The data shows a total of ‘r nrow(chlorine)‘ chlorine results.`

```
# Conclusion and Recommendation
- Chlorine levels in some Gormsey suburbs are high
- High levels lead to a negative customer experience
- Highest level in ‘r chlorine_results$Suburb[which.max(chlorine_results$Taste)]‘.
- Further research required to correlate taste complaints with chlorine levels
```

6.6.6   Knitting the Document

When R knits the document, it first evaluates all code and converts the output to Markdown format. The Markdown file is then passed to the pandoc software bundled with RStudio that can convert it to the required output format. Pandoc is separate software that can convert many document types and comes bundled with RStudio.

When working on a project, it is best to first write the code in a well-commented R script and copy it into a Markdown document after completing the analysis. You can then add explanatory text to create reproducible research.

6.7   Presenting Numbers

The numerical output of R functions often contains too many decimals. Several functions are available to control the way R presents numbers. Firstly, you can set the default number of digits with the `options(digits = n)` function. Standard R is accurate up to about 15 decimals. These limitations are caused by the hardware.

If you need more decimals, you must use specialised packages, such as *Rmpfr* (Maechler, 2023). This package provides multiple precision arithmetic, using software instead of hardware to calculate large numbers with lots of decimals. The precision of this method is only limited by the amount of memory available in your computer.

The `round()` function defaults to round to an integer or a defined number of decimals. For example, when using a negative number in the `digits` option, the number is rounded to the nearest power of ten. Wrapping an expression between parentheses will print it in the console.

```
(a <- sqrt(777))

[1] 27.87472

round(a) # Show as integer

[1] 28

round(a, digits = 2) # Round to two digits

[1] 27.87

round(a, digits = -1) # Rounding to a power of ten

[1] 30
```

The `floor()` and `ceiling()` functions result in an integer.

```
floor(a) # Remove all digits
```

```
[1] 27
```

```
ceiling(a) # Round up to the nearest integer
```

```
[1] 28
```

If you like to constrain the number of total digits, including the integer part, use the `signif()` function.

```
signif(a, 5) # Show 5 characters
```

```
[1] 27.875
```

More advanced formatting of numbers is available through the `format()` function. The format function has many options to change how you display numbers. The most common option is `big.mark = ", "`, which adds a comma to separate thousands. The `scientific` option toggles scientific notation on and off.

```
format(2^32, big.mark =",")
```

```
[1] "4,294,967,296"
```

```
format(2^128, big.mark =",")
```

```
[1] "3.402824e+38"
```

```
format(2^128, big.mark =",", scientific = FALSE)
```

```
[1] "340,282,366,920,938,463,463,374,607,431,768,211,456"
```

6.7.1 Pasting Results Together

Another useful function is `paste()`, which concatenates strings and variables. By default, it adds space as a separator between the elements, which you can change with the `sep` parameter. For example:

```
paste("Mean THM value is", round(mean(thm$Result), 2), "Mg/l", sep = ": ")
```

```
[1] "Mean THM value is: 0.04: Mg/l"
```

6.8 Further Study

The project folder contains an R Markdown file (`chlorine-taste.Rmd`) in the `course` folder with an example using the Gormsey data. This file analyses a problem and presents the analysis as a slide deck. One of the slides uses a two-column layout. To achieve this in Markdown, you need to use the slightly cumbersome format shown below:

```
:::::::{.columns}
:::{.column}
Text / code goes here
:::
:::{.column}
Text / code goes here
:::
:::::::
```

This chapter shows that Markdown is a powerful software that allows you to combine code with text. In addition, several packages exist that enhance the capability of R Markdown, such as the *bookdown* package, which helps you to write professionally-designed books (Xie, 2017).

Writing in plain text has many advantages over using text editors. Many journals now expect authors to use literate programming to improve the peer-review process. R Markdown is not the only tool for literate programming with the R language. This book is written with Emacs and Org Mode, which is versatile software for advanced users. Some scientists also use LaTeX and Sweave to combine text with code to write papers and books.

The most advanced method to create reproducible research is developing a dynamic dashboard with the *Shiny* package. This package can create dynamic applications you can share through a web browser (Sievert, 2020).

7

Managing Dirty Data

The data used in the first case study is perfect in many ways. There are no extra columns or missing observations, and everything is perfectly labelled and ready to be analysed. Data in the real world is, however, not always this clean. Therefore, preparing data for analysis, sometimes called data munging or wrangling, is an essential part of the data science workflow. This chapter introduces some techniques to clean data with R and the Tidyverse to create reproducible code.

The case study for the next four chapters is a survey that customers in our imaginary city of Gormsey completed. Analysing customer surveys requires different techniques to analyse data compared to physical processes, and this case helps technical professionals apply the principles of psychometric analysis. The learning objectives for this session are:

- Use the *dplyr* package to transform data

- Apply the principles of tidy data

- Develop a script to automate data cleaning

7.1 Case Study 2: Understanding the Customer Experience

The Gormsey water utility decided that it would be good to know how the people of Gormsey feel about their water services. The customer experience manager has three research questions:

1. How much do people care about the service by the Gormsey water utility (Chapter 8)?
2. What is the relationship between contact frequency and financial hardship segments (Chapter 9)?
3. What customer segments can we find in the data (Chapter 10)?

The sample frame for this survey are all customers of the Gormsey water utility. A random sample of consumers from three suburbs completed a series of questions. The data used in this case study is a modified version of data collected for a PhD research project about customer-centricity for water utilities (Prevos, 2017). The survey contains information about:

- Customer demographics

- Financial hardship

- Contact frequency

- Involvement with tap water

- Service quality

Our task is to analyse the results of this survey to answer these questions. But we need to clean the raw data before answering these questions.

7.2   Cleaning Data

The second step in the data science workflow visualised in Figure 6.1 is to clean or tidy the data. This phase involves transforming the data as it is collected and stored into a format suitable for analysis.

The literature and websites about data science often claim that cleaning data can consume eighty percent of the available time. Although this often-cited statement has no empirical bases, cleaning is a fundamental part of the data workflow and can take a frustratingly long time. Furthermore, cleaning data is essential because even the most advanced algorithm cannot create value from dirty data. As the old saying goes: "Garbage in is Garbage Out" (GIGO).

The main benefit of cleaning the data with code is that the process is reproducible and replicable. The process is reproducible because the original raw data does not change, and all steps are documented. The process can be rolled back or corrected when the need arises. Using code to clean data is also replicable because it can be repeated with other raw data sets with the same structure, for example, when repeating the same survey. Cleaning data generally involves three steps:

1.   Load and explore the data

2.   Convert the data structure

3.   Remove invalid data

7.3   Load and Explore the Data

The survey results are stored in the `data` folder in the `customer_survey.csv` file. This file contains the raw data collected from the online survey system. Section 2.3.1 describes how to obtain this data.

```
library(readr)
library(dplyr)
rawdata  <- read_csv("data/customer_survey.csv")
glimpse(rawdata)
```

Using `rawdata` as a variable name keeps this data intact as we process it when we need to use it again. This practice prevents having to reload the data every time you change the script. However, loading data in real life can be time-consuming and requires lots of computing resources, so you want to minimise the number of times you do so.

To determine the steps to clean this data, we need to work backwards and define what our data need to look like after cleaning. The survey in this case study measured six areas of interest using Likert and Semantic Differential Scales. These are survey techniques where a respondent selects a statement that most closely matches their preference, such as "Totally Disagree" or "Totally Agree". These responses are converted to numbers to enable statistical analysis. To enable this analysis, we thus need a data frame that contains these numeric values for each variable. It is also good data management practice that each row has a unique identifier. So our clean data should look like this:

• Unique respondent ID (`V1`)

• Customer suburb (`suburb`)

• Customer involvement (ten items: `p01` – `p10`)

- Hardship (single item: `hardship`)

- Contact frequency (single item: `contact`)

- Technical service quality (five items: `t01 – t05`)

- Functional service quality (thirteen items: `f01 – f13`)

7.4 Convert the Data Structure

The second step in our cleaning process is transforming the data into the required format and structure. All variables need to be of the correct class (number, character, date, and so on) and only keep those variables needed for analysis. We will also need to join some metadata to add context to the responses.

The output of the `glimpse()` function shows that this data contains over fifty variables. The first eighteen columns contain metadata, such as a unique response ID, internet addresses, start, and end times, and so on. The following columns contain the actual responses. Also of interest is that the class of all variables in the raw data is a character, even though the survey responses are numbers.

7.4.1 Convert Data Types

Looking at the data with the `View()` or `glimpse()` functions shows that the first two rows contain header information. The header names look fine, but the first row of data consists of text, which is why R assigned all variables as characters. A clean data set should have only one header row, so we need to remove this first row. We could skip the first row with the `skip = 1` parameter in the `read_file()` function, but the second line in the CSV file (first row in the data) needs to be removed.

```
customers <- rawdata[-1, ]
customers <- type_convert(customers)
```

The first line of code creates the new `customers` data frame by removing the first line of the raw data. Using negative numbers in data frame indices removes them from the output. Next, the `type_convert()` function re-assesses the data frame to guess the correct data classes. Using the `glimpse()` function again, we can see that most columns are now numerical values `<dbl>`, which is what we want them to be. This function can convert a data object to logical, integer, numeric, complex, character, or factor.

7.4.2 Select Relevant Variables

The survey data contains metadata that we don't need for further analysis. The first columns contain information about when the survey was taken and so on. We don't need this data for further analysis, except the `term` variable, discussed in Section 7.5.

The `names()` function lists the names of the variables in a data frame and helps us to select the ones we like to keep. The `select()` function in the *dplyr* package works like the filter function, but for columns. You can use numbers or names to indicate the required columns (negative numbers remove a column). In this case, we like to keep the first column, which is the unique ID for each respondent (column 1), and the columns with the survey responses (columns 19–51).

```
names(customers)
customers <- select(customers, c(1, 15, 19, 21:51, -33))
```

We should also rename the V1 to customer_id because that is a bit more descriptive. The names of the other variables (p01 etc.) seem cryptic, but they will be explained later. As the name suggests, the rename() function from the *dplyr* package changes the name of variables in a data frame.

```
customers <- rename(customers, customer_id = V1)
```

7.4.3   Joining Data Frames

The data is starting to look much better, but there are a few more steps before we reach the ideal state. The suburb is just a number, which ideally should be a name. We can achieve this by merging the data with a dimension table. In data architecture, fact tables contain the observations (such as the customers data frame), while dimension tables contain contextual information. Dimension tables are often needed in surveys with drop-down boxes to provide answers, as the data is usually stored as numbers. Using this data approach saves a lot of storage space, as you only hold a number instead of repeated character strings, but it is not suitable for reporting. The dimension table for this survey contains the relationship between the numbers and the names of the suburbs. In this case study, the three surveyed suburbs are:

1. Merton
2. Tarnstead
3. Wakefield

First, we create the dimension table to link the numbers with suburbs, which is then joined to the fact table, after we can ditch the city variable. The minus sign in the select() function removes a column.

```
suburbs_dim <- tibble(suburb = 1:3,
                      suburb_name = c("Merton", "Tarnstead", "Wakefield"))

customers <- left_join(customers, suburbs_dim)
customers <- select(customers, -"suburb")
```

The left_join() function finds the matching fields in the two sets and then merges the data frames. Since both sets have a variable named suburb of the same type, the join function automatically matches these fields. A left_join() returns all rows from the left tibble (customers) and all columns from both tibbles (customers and cities). If one of the customer data's city variables contains a number not in the dimension table, the suburb_name variable becomes NA. If the dimension table has missing matching references, the result is also NA. If there are multiple matches, all combinations of the matches are returned.

When performing the join action, R will display the names of the fields the data was joined on in the console, in this case, suburb. If the two tables had different field names for the city, then you use the by parameter. If for example, the dimension table used suburb_num, you would use: left_join(customers, suburbs_dim, by = c("suburb" = "suburb_num")).

The left_join() function is the most common way to join two data sets. The Tidyverse has several other join functions that match values differently. The Tidyverse *_join() functions merge two data frames or tibble into one. The R base package uses the merge() function for this purpose, but it does not have the flexibility of the Tidyverse equivalents. When you need to join two data frames together, you can use the rbind() and cbind() functions. These functions can bind two data frames by appending rows or columns. These functions only works when the joined data frames are compatible in size and variable type. The Tidyverse versions of these functions are bind_rows() and bind_cols(), which are a bit more flexible when it comes to the structure of the merged data frames.

7.5 Remove Invalid Data

Not all data is created equal; an essential cleaning step is removing invalid responses. The definition of invalid data depends on the type of information and its context. Data collected through surveys is rarely perfect. Respondents might not complete all questions, not pay attention, or are not inside the sample frame. The next step is thus to remove any respondents who did not meet specific criteria:

- Didn't consent to their data being used

- Don't have tap water

- Don't live in one of the three nominated suburbs

- Didn't pay attention when completing the survey

The first question on the survey asks for consent to use the responses. Informed consent is the first principle of ethical data science (Section 1.4.4). Respondents who did not consent were exited from the survey, so their responses are missing.

The sample frame for this survey are people living in three suburbs of Gormsey who have tap water. To test whether a respondent should be in the sample, the survey asks whether the respondent has tap water and which suburb they live in.

This survey paid respondents to complete the questions using a survey panel. One of the problems with using paid subjects is that they are motivated to complete many surveys without having much regard for their answers. Respondents were therefore subjected to an attention filter: "If you live in Gormsey, select Strongly Agree". The survey was only sent to people within Gormsey. Respondents who did not answer "Strongly Agree" should, as such, be excluded from the sample because they did not pay attention.

The `term` field reports why respondents terminated the survey and whether they paid attention. The `table()` function helps to summarise the content of this field.

```
table(customers$term)
```

```
attention    consent noTapWater  otherCity
       79          8         15         97
```

The output of this function shows that 79 people did not pass through the attention filter, 8 did not consent, 15 did not have tap water, and 97 lived outside Gormsey. You might notice that the total number of items in the table does not match the number of rows (observations) in the data frame. This is because when you view the content of the customers$term field, you see many entries with NA in them. These are empty values (Not Available). R uses this method to manage missing observations. The table function can include NA values with the =useNA = "ifany" = option, which reveals that 491 respondents in our sample frame properly completed the survey.

```
table(customers$term, useNA = "ifany")
```

```
attention    consent noTapWater  otherCity      <NA>
       79          8         15         97       491
```

We only want those rows of data with an NA value in the term field, as these are the surveys that were not terminated. To find these observations, we need to use the is.na() function. The function results in a logical variable (TRUE or FALSE that shows whether a field is not available. You cannot use term = NA= because missing data is a special condition.

The following line of code filters the customer data to include only those surveys that were not terminated. These are the people who consented and fit in the sample frame. We can then remove the term variable, as it has served its purpose.

```
customers <- filter(customers, is.na(term))
customers <- select(customers, -"term")
```

To see all respondents who terminated the survey, you can negate the is.na(term) statement with an exclamation mark (!is.na(term)). We are thus asking R to filter the customer data frame for all entries where termination is not unavailable (NA). You gotta love double negatives.

Besides clean data, we also need clean code to ensure that the process of cleaning and analysing data can be understood by others who like to replicate what you did. Therefore, this chapter will also show how to write code that is easy to read and understand.

7.6   Refactoring Code

Chapter 2 made some references to coding style to ensure it is readable not just to you but also to other people. Besides clean data, we therefore also need clean code. The previous paragraphs contain a lot of code developed on the fly. Refactoring is making the code more efficient and easier to read.

The sequence of functions explained above cleans the data for further analysis from the raw data to the finished product. The benefit of this approach is that the raw data remains unchanged, so we can also use this code on other survey results with the same data structure. However, the code is repetitive because we change the customers variable several times in a sequence. In summary, we have taken the following steps to clean the data (column selection is moved to the end, so it is only used once):

```
customers <- rawdata[-1, ]
customers <- type_convert(customers)
customers <- select(customers, c(1, 15, 19, 21:51, -33))
customers <- rename(customers, customer_id = V1)
customers <- left_join(customers, suburbs_dim)
customers <- select(customers, -"suburb")
customers <- filter(customers, is.na(term))
customers <- select(customers, -"term")
```

A rule of thumb in R coding is that if you repeat the same thing more than twice, there will be a more efficient way of achieving the same result. In this example, we used "customer <-" eight times, and the select function appeared three times. There are two ways of combining these lines of code to reduce repetition.

A typical way to code in a spreadsheet is to join the steps with nested functions. While the nested approach uses less space, it is not as easy to understand because you have to read from the inside out. This code combines the three select() function calls and only one declaration of the customers variable.

```
customers <- rename(
  select(
    rename(
      left_join(
        filter(
          type_convert(rawdata[-1, ]),
          is.na(term)),
        suburbs_dim),
      customer_id = V1),
    c(1, 52, 21:51, -33)))
```

7.6.1 Using Tidyverse Pipes

The *magrittr* package within the *Tidyverse* uses the pipe operator to streamline this process (Bache & Wickham, 2022). A pipe transports the output of one function to the input of the next one. A pipe replaces `f(x)` with `x %>% f()`, where `%>%` is the pipe-operator. This operator works almost the same as the plus symbol in the *ggplot2* package (Chapter 5). This code means that R pipes the value of `x` to the function `f()`. This step can be repeated in a long sequence. You can quickly type this operator with the `Control-Shift-M` keyboard shortcut. The code used to clean the customer data is now written like this:

```
customers <- rawdata[-1, ] %>%
  type_convert() %>%
  filter(is.na(term)) %>%
  left_join(suburbs_dim) %>%
  rename(customer_id = V1) %>%
  select(c(1, 52, 21:51, -33))
```

You need to read this code sequence as follows: The raw data without the first row is piped to the type converter. The output from this step goes to the filter, onward to the join function, and so on. The name of the `customers` variable only appears once because it is transported through the pipe. The pipe operator moves the output of the previous step to the first parameters in the following function. You don't write the first parameter for each function because it is implied in the pipe. The best way to understand this piped code is to evaluate each step and review the output. You can select bits of the code and run them separately. Make sure you don't include a pipe, as R will otherwise wait for further input.

We now have a reproducible script that can be reused every time we run this survey. This approach promotes the reproducibility of the analysis and allows for peer review of the investigation.

7.6.2 Data Cleaning Workflow

The final step saves this data to disk so we can reuse it in the following chapters. Changing the content of a data frame only changes it in memory and not on disk. When you close RStudio, it will ask you to save the data in a backup file. R will not change the original CSV file unless you instruct it. For example, the `write_csv()` function takes a data frame and saves it as a CSV file to the specified filename. Best practice is to not change the raw data and use a different name. Keeping the raw data intact means that your code is reproducible, and you can always go back to the original state of the data.

```
write_csv(customers, "data/customer_survey_clean.csv")
```

Writing the data to disk is a suitable method in cases where the raw data will not change. Another method is to run the cleaning code every time you run the analysis. This method ensures that the analysis uses the latest changes in the source data or cleaning method. The refactored code for this chapter is:

```
## Cleaning customer survey data
library(readr)
library(dplyr)

# Suburb dimension table
suburbs_dim <- tibble(suburb = 1:3,
                      suburb_name = c("Merton",
                                      "Tarnstead",
                                      "Wakefield"))

# Clean data
customers <- read_csv("data/customer_survey.csv")[-1, ] %>%
  type_convert() %>%
  filter(is.na(term)) %>%
  left_join(suburbs_dim) %>%
  rename(customer_id = V1) %>%
  select(c(1, 52, 21:51, -33))

# Remove helper variable from memory
rm(suburbs_dim)
```

You can save this script and reuse it within other scripts and then call it with the `source()` function. Then, every time you run this script from within another script, the `customers` data frame is recreated from scratch, ensuring you always use the latest data.

```
source("scripts/07-customers_clean.R")
```

The data is now ready for analysis, but there are still some issues. The survey measures different aspects. The questions that start with t measure technical service quality. This construct is a measure of how customers experience the core services of a water utility. The remainder of this chapter investigates these five variables:

- t01: Tap water is available whenever I need it

- t02: My tap water is always safe to drink

- t03: My tap water is always visually appealing

- t04: My tap water always has a pleasant taste

- t05: My tap water always has sufficient pressure

7.7 Dealing with Missing Data

The reality of measurement is that we don't always have a complete record, whether there are issues with instrumentation or telemetry, or respondents did not complete all questions. We thus have to deal with missing data points. While electronic surveys can make answers compulsory, this strategy will also increase the number of respondents who drop out, resulting in a smaller sample. To review the completeness of the data, we can use the `summary()` function for the variables from t01 to t05.

```
      t01              t02              t03              t04              t05
 Min.   :1.00    Min.   :1.00    Min.   :1.000   Min.   :1.000   Min.   :1.000
 1st Qu.:6.00    1st Qu.:4.00    1st Qu.:5.000   1st Qu.:4.000   1st Qu.:5.000
 Median :7.00    Median :6.00    Median :6.000   Median :5.000   Median :6.000
 Mean   :6.25    Mean   :5.35    Mean   :5.475   Mean   :4.956   Mean   :5.576
 3rd Qu.:7.00    3rd Qu.:7.00    3rd Qu.:7.000   3rd Qu.:6.000   3rd Qu.:7.000
 Max.   :7.00    Max.   :7.00    Max.   :7.000   Max.   :7.000   Max.   :7.000
 NA's   :59      NA's   :59      NA's   :59      NA's   :59      NA's   :59
```

When you study the output, you see that for all ten variables, the minimum is 1, and the maximum is 7. At the bottom of the result for each variable, you will also note that there are 59 missing responses, which R indicates with NA (Not Available).

Missing data can be random or through an underlying pattern. We need to deal with each type of missing data differently. Data *Completely Missing At Random* (CMAR) is part of the sampling error. The missing data is independent of any other variables in the survey. Randomly missing data can be either ignored or imputed (replaced with expected values). Depending on your type of analysis, you might have to remove respondents with random missing data from the sample. Data that is *Missing Not At Random* (MNAR) indicates an underlying pattern. These respondents are, in most cases, omitted from the analysis.

The fact that the same data points are missing for each variable is an intriguing clue. The data in this survey seems to be MNAR. Reviewing the data shows that 52 respondents did not answer the involvement questions. When data is MNAR, as in this case, we usually need to remove these observations, which we will do in the next step.

The *visdat* package provides a visual interface to explore missing data (Tierney, 2017), as shown in Figure 7.1. You can read the vignette to explore this package: `vignette("using_visdat")`.

```
library(visdat)
vis_miss(tq)
```

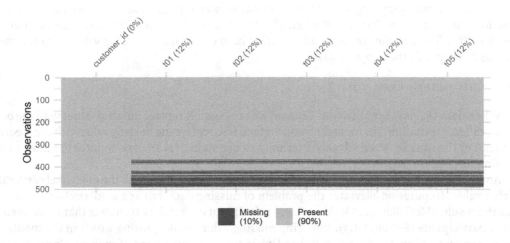

FIGURE 7.1 Missing survey responses.

Collecting data is not a perfect process. Survey respondents might not answer all questions, or technical issues prevent completeness in the case of measurements. Dealing with missing data can be frustrating as it reduces the statistical power of your analysis as the sample size effectively reduces. You can do three things with missing data:

1. Calculate with missing data
2. Remove missing data from the sample
3. Impute data by filling it with expected values

7.7.1   Calculating with Missing Data

Using missing data requires special considerations during analysis. Almost all functions will return an NA value when one or more observations are unavailable, as shown in the example below. Most functions accept the na.rm = TRUE option to instruct R how to deal with missing values. The second line of code tells R to remove any NA values from the vector. The default setting for this option is to keep the missing observations, which can be frustrating but protects you from drawing wrong conclusions. The help file for each function will include instructions on how to deal with missing data.

```
mean(tq$t01)
```

```
[1] NA
```

```
mean(tq$t01, na.rm = TRUE)
```

```
[1] 6.25
```

Another method is to remove any missing values from a vector with the na.omit() function.

```
mean(na.omit(tq$t01))
```

```
[1] 6.25
```

7.7.2   Remove and Impute Missing Data

This analysis removes any respondents who did not answer any questions. The complete.cases() function creates a vector that is TRUE when all values in a row are available and FALSE when data is missing. This function thus removes all rows that have one or more missing values. We can then use this to keep only the complete cases.

```
tq <- tq[complete.cases(tq), ]
```

When data is *Completely Missing At Random*, we can possibly replace missing values with the best guess, called imputation. The most common method is to replace the missing value with the median or mean of the sample. More advanced techniques use statistical analysis to infer the most likely missing response.

Another approach is to impute data, which replaces missing data with the most likely estimate of that value. Imputation alleviates the problem of missing data, but at a statistical cost as it can bias the results (McCallum, 2013). The second principle of ethical data science is that we do justice to the participants (Section 1.4.4). Imputing missing values is like putting words in the mouth of the respondent. Imputation can only be used when the primary method of analysis cannot process missing values and when the number of missing values is only a few percent of the total number of observations. Imputation should never be used to increase the sample size.

7.8 Tidy Data

We cleaned the survey data in the previous session by removing unwanted columns and respondents. Although the data is clean, it is not yet in its ideal tidy state. Tidy data is a standard way of mapping the meaning of a data set to its structure. Data that is structured in a tidy way is more natural to analyse and visualise (Wickham, 2014). A data set is a collection of values, primarily numbers or strings of characters. Every value belongs to a variable (column) and observation (rows). A variable should contain all values that measure the same underlying attribute (like height, temperature, duration). An observation provides all values measured on the same unit (like a person, day, or location) across attributes. Data is messy or tidy depending on how rows, columns, and tables are matched up with observations, variables, and types. In tidy data:

- Each variable forms a column

- Each observation forms a row

- Each type of observational unit forms a table

The laboratory results sets used in the first case study are tidy because all measurements are in the same result column (Section 3.4). However, the customer data is untidy because each respondent's results are spread across ten columns. This data structure is more challenging because it cannot be grouped. The functionalities of the Tidyverse work best with tidy data. However, other functions require a wide format, as we shall see in the following chapters.

To tidy the data, we need to pivot the data to a longer format. The `pivot_longer()` function in the *tidyr* package helps to create tidy data. This function takes multiple columns and collapses them into key-value pairs. The example in Figure 7.2 transforms a wide version into a tidy long version.

The first option in the `pivot_longer()` function is the data frame's name to be transformed. The next option defines which columns need to pivot, which are the ones that contain the data. The remainder of the columns will be used as the keys. The last two options provide the names of the new columns. The `names_to` option defines the column name that will store the names of the pivoted variables. The `values_to` option specifies the name of the column that will hold the values in the pivoted columns: `pivot_longer(data, -Measure, names_to = "Location", values_to = "Result")`.

Measure	SP1027	SP1284
Chlorine	0.794	0.918
THM	0.001	0.012
Turbidity	0.125	0.701

Measure	Location	Result
Chlorine	SP1027	0.794
THM	SP1027	0.001
Turbidity	SP1027	0.125
Chlorine	SP1284	0.918
THM	SP1284	0.012
Turbidity	SP1284	0.701

FIGURE 7.2 Principles of the pivot longer and wider functions.

For the customer data, we include all columns, except the respondent `customer_id` and suburb. The names column will be called `Item`, which contains the name of the thirty items. The values column is `Response`, which contains the scores.

```
library(tidyr)
tq_long <- pivot_longer(tq, cols = -1,
                        names_to = "Item",
                        values_to = "Response")
```

With this tidy data set, we can visualise some groups of responses. The graph in Figure 7.3 visualises the technical service quality responses. Note how the `scale_y_continuous()` function configures the *y*-scale to display all seven numbers (`1:7`). Using `scale_y_continuous(breaks = c(1, 7))` would only show numbers 1 and 7. The `scale_x` and `scale_y` layers have extensive capabilities to manipulate the look and feel of the plot axes.

```
library(ggplot2)

ggplot(tq_long, aes(Item, Response)) +
   geom_boxplot() +
   scale_y_continuous(breaks = 1:7) +
   labs(title = "Technical Service Quality",
        subtitle = "Tap water in Gormsey")
```

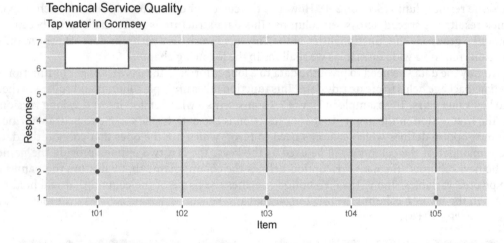

FIGURE 7.3 Technical service quality items.

We now have our first result for the second case study, which seems to suggest that customers are content with the technical quality of the water services in Gormsey. The following chapters will dig deeper into this case study to validate the survey results and uncover some patterns.

7.9   Further Study

This chapter shows how to clean the case study data using reproducible code. The code in this chapter is only an example, as each data set will require bespoke cleaning steps. The first step is to explore the raw data, then define the desired end state and write the appropriate code to get to that state.

You have now seen quite a few of the functions, also called verbs, of the *dplyr* package. The ones we have used so far are:

- `count()` counts the number of rows in each group

- `filter()` picks cases based on their values

- `group_by` performs operations by grouped variables

- `summarise()` reduces multiple values down to a summary

- `select()` picks variables based on their names

- `mutate()` adds new variables that are functions of existing variables

- `rename()` changes the name of a variable

- `left_join()` joins two tables

These functions provide a powerful toolkit to transform and analyse data. These are, however, not the only functions in this package. The *dplyr* vignette (`vignette("dplyr")`) describes all verbs available in this powerful package. Read this page to familiarise yourself with the capabilities of this package.

8

Analysing the Customer Experience

The analysis in the previous chapters describes data using means, medians, and other such statistics and graphs. Relying on descriptive statistics (Chapter 4), as numbers or visualisation, is like trying to get to know somebody by describing their physical appearance. If I tell you that John is a tall man with dark greying hair, that does not tell you anything about who he is. Summarising data descriptively does not add new knowledge but merely summarises what you already know. The remainder of the book discusses inferential statistics, which is a set of methods for finding patterns in data to create new knowledge.

This chapter starts with the second case study in which we analyse data from customer surveys. Water professionals often focus on the tangible aspects of water services and measure performance in cubic metres, gallons, or kilolitres. Technical data does, however, only tell part of the story of urban water supply. We can physically measure the process from catchment to tap, but what happens downstream of the connection is a matter of psychology and sociology instead of chemistry and physics. This chapter discusses how to use surveys to learn how customers experience water services and demonstrates some techniques to analyse this type of data using the R language. The learning objectives for this chapter are:

- Understand the principles of measuring the customer experience with surveys

- Assess the reliability of surveys with a correlation matrix

- Assess the validity of surveys using factor analysis

8.1 Measuring Mental States

Technical professionals often lament that customer data is merely subjective and that it, therefore, cannot provide real insights. The following three chapters demonstrate some techniques social scientists use to construct and analyse surveys. While each answer is a subjective assessment, a well-designed and appropriately analysed customer survey can provide actionable insights into how a utility can improve its services perceived by the customer.

While we can measure physical aspects of water and wastewater directly, the customer experience is much harder to capture in numbers. Marketers, psychologists, and sociologists have developed advanced techniques to use surveys to learn about our state of mind. Unlike physical measurements in a water treatment plant, the state of mind of a consumer cannot be measured directly. Even the most advanced brain scanning techniques are unable to determine how satisfied a customer is when using a product or any of the other phenomena we might want to like to investigate.

Surveys are the most common method to measure psychological constructs such as satisfaction. Water utilities have, in recent years, extensively used surveys to better understand their customers and to measure the level of service downstream of the water meter. However, developing surveys requires expertise and careful planning to ensure that the data provides the needed insights (Goetz, 2016).

The underlying assumption of customer surveys as a measurement tool is that a causal relationship exists between the respondent's state of mind and the answers they provide on the survey. However, measuring a state of mind is a complex task beyond asking direct questions. For example, simply asking: "How satisfied are you with tap water?" would not yield necessarily reliable and valid results. Firstly, this question assumes that the respondent has the same understanding of satisfaction as the researcher. Secondly, with only one question, there is insufficient information to test the reliability and validity of the responses, and we have to take the answer at face value.

Physical measurements can be calibrated by comparing them with a known value. For example, we can compare a length measurement with a known standardised value, calibrate a flow meter by measuring the volume pumped over a fixed time interval, and so on. Psychological states of mind cannot be calibrated as we have no direct insight into the software of the mind. In psychology and the social sciences, mental states are modelled as latent variables because we can only measure them indirectly (Figure 8.1).

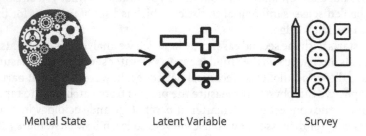

Mental State   Latent Variable   Survey

FIGURE 8.1 Relationship between mental states, latent variables, and survey questions.

Researchers use banks of questions that ask for a response to similar items. The words in the survey item are a stimulus that solicits a known response from the mind. The basic idea is that people with a similar disposition will respond similarly to these stimuli. We then use the data to test the responses for internal and external consistency. Measuring customer satisfaction, for example, involves asking respondents three or more questions about topics that contribute to satisfaction, such as friendliness of employees, clarity of their bill, and the taste of water.

Marketing researchers, sociologists, and psychologists have published statistically validated survey scales that can measure these latent variables, such as personality, consumer trust, service quality, and many more (Bruner, 2012; Bearden, 2011).

8.1.1 Reliability and Validity

As discussed in Section 1.4.2, data science needs to be valid and reliable. Reliability means that the responses have acceptable accuracy and consistency across samples. Validity in this sense means that a survey measures the psychological construct we seek to understand. Using an archery analogy, reliability means our arrows (measurements) are close, and validity means we hit the bullseye (Figure 8.2). A outcome or a measurement is valid and reliable if it provide a narrow band of possible options and is centred around the true value.

The reliability of physical or psychological measurements depends on the quality of the instrumentation used to obtain the data. Engineers spend much effort to ensure that instruments are reliable through maintenance and calibration programs. In addition, the instruments need to be installed and maintained to the manufacturer's specifications to ensure their ongoing accuracy. The same principles apply to developing surveys to measure how customers think and feel about tap water services. Surveys need to be well-designed and calibrated using statistics.

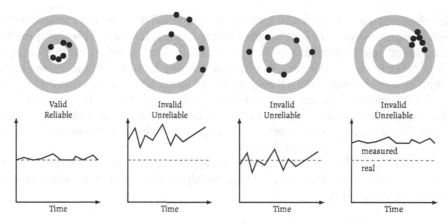

FIGURE 8.2 Validity and reliability of data and analysis.

The remainder of this chapter explains how to use statistical analysis to assess the reliability and validity of customer surveys using the data from the second case study, so we start with the data cleaning script developed in Section 7.6.2).

```
source("scripts/customer_clean.R")
```

8.2 Case Study: Consumer Involvement

Consumer involvement is an essential marketing metric that describes a product's or service's relevance in somebody's life. For example, people who own a car will be highly involved with purchasing and owning the vehicle due to the large amount of money involved and the social role it plays in developing their public self. Conversely, consumers have a much lower level of involvement with the instant coffee they drink than with their clothes. More formally, consumer involvement can be defined as a person's perceived relevance of a good or service based on inherent needs, values, and interests.

Understanding involvement in the context of urban water supply is essential because sustainably managing water as a common pool resource requires the active involvement of all users. The lowest level of involvement is a state of inertia, which occurs when people habitually purchase or consume a product without further thought. Water is available in the background of everyday life, which suggests a low level of involvement. However, the essential nature of water indicates a high level of involvement. This survey measures the involvement construct to better understand how involved consumers are with their water service.

8.2.1 Personal Involvement Inventory

The customer survey of the second case study includes ten questions to measure the level of consumer involvement. These questions form the *Personal Involvement Inventory* (PII), developed by Judith Zaichkowsky (1994). The inventory has two dimensions:

1. Cognitive involvement (importance, relevance, meaning, value, and need)

2. Affective involvement (involvement, fascination, appeal, excitement, and interest).

The involvement question bank uses a semantic differential scale. This method requires respondents to choose on a seven-point scale between two antonyms (Table 8.1). This survey method measures the meaning people attach to a concept, such as a product or service. The items are presented in a random order to each respondent. In principle, the words on the right indicate a high level of involvement. However, five questions have a reversed polarity, which means that the left side shows a high level of involvement. For example, *important* implies a high level of involvement but is at the start of the scale, while *worthless* implies a low level of involvement. The following section shows how to normalise these responses. Reversed polarity is a technique to ensure respondents remain attentive by reversing the direction of the scale.

The customer survey data we cleaned in the previous chapter contains the ten items of the PII scale (p01, p02 ... p10). Table 8.1 shows the relationship between the items and the scale, some of which are in reversed polarity.

TABLE 8.1 Personal Involvement Index questions.

Variable	Item	Polarity
p01	Important – Unimportant	Reverse
p02	Relevant – Irrelevant	Reverse
p03	Meaningless – Meaningful	Normal
p04	Worthless – Valuable	Normal
p05	Not needed – Needed	Normal
p06	Boring – Interesting	Normal
p07	Exciting – Unexciting	Reverse
p08	Appealing – Unappealing	Reverse
p09	Fascinating – Mundane	Reverse
p10	Involving– Uninvolving	Reverse

8.2.2   Preparing the Involvement Data

To analyse the level of involvement, we only need the customer_id as a unique identifier and the ten PII items. The starts_with() helper function lets you choose columns based on a prefix. But before we can analyse the variables of interest, we must correct the reversed polarity. We can reverse the five revered items by subtracting the response from eight. The dplyr mutate() function changes variables or creates new ones within a data frame.

```
pii <- select(customers, customer_id, starts_with("p")) %>%
  mutate(p01 = 8 - p01,
         p02 = 8 - p02,
         p07 = 8 - p07,
         p08 = 8 - p08,
         p09 = 8 - p09,
         p10 = 8 - p10)
```

In the code used to correct polarity, the mutate() function acts on individual variables in columns, which means you have to repeat the same action for each column. The *dplyr* package can also mutate variables over multiple columns with the mutate_at() function. Note that in this function, the variable names have to be quoted. You nest the data transformation in a function, where x (or any other variable name of choice) becomes the value of the indicated variables. Functions are explained in more detail in Chapter 12. In this code snippet, those pesky missing values (see Section 7.7) are removed by only using the rows with complete data.

```
pii <- select(customers, customer_id, p01:p10) %>%
  mutate_at(c("p01", "p02", "p07", "p08", "p09", "p10"),
            function(p) 8 - p)

pii <- pii[complete.cases(pii), ]
```

To visualise the data, we need to pivot the data to the long format so that *ggplot2* can group the items. Visualising each survey item with a boxplot provides a quick insight into their distributions. The visualisation also shows an interesting pattern.

```
library(ggplot2)
library(tidyr) # For the pivoting function

pii %>%
  pivot_longer(-customer_id, names_to = "Item", values_to = "Response") %>%
  ggplot(aes(Item, Response)) +
  geom_boxplot() +
  theme_minimal()
```

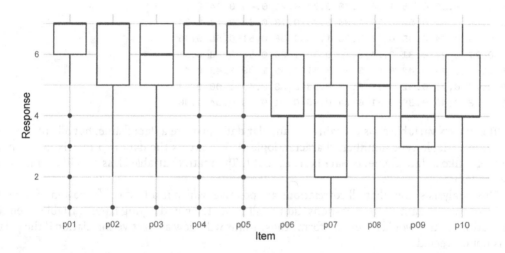

FIGURE 8.3 Personal Involvement Index responses.

The image in Figure 8.3 shows an interesting pattern that confirms the theoretical assumption of the Personal Involvement Index. The first five questions (cognitive involvement) all score much higher than the last five items (affective involvement). Before drawing any conclusions about this pattern, we need to verify whether these results are reliable and valid.

8.3 Measuring Reliability

8.3.1 Correlation between Responses

The first step in assessing the reliability of the involvement responses is to review the correlations between the items. If these questions do indeed all relate to an underlying latent variable, then the responses should correlate strongly with each other.

The cor() function calculates the correlation between two vectors. The output of this function is a single number, and the correlation between p01 and p02 is approximately 0.66.

```
cor(pii$p01, pii$p02)
```

```
[1] 0.6581375
```

We could do the same action for all 45 unique combinations of variables, but that would be tedious. The correlation function can also analyse a whole data frame at once. In this case, we need to remove the first column because it is not a numerical variable. The output of this code is a matrix of correlations between all combinations of p01 to p10, rounded to two decimals.

```
c_matrix <- cor(pii[, -1])
round(c_matrix, 2)
```

```
      p01  p02  p03  p04  p05  p06  p07  p08  p09  p10
p01  1.00 0.66 0.45 0.62 0.54 0.23 0.27 0.40 0.24 0.28
p02  0.66 1.00 0.49 0.56 0.51 0.24 0.27 0.44 0.27 0.40
p03  0.45 0.49 1.00 0.55 0.52 0.35 0.31 0.42 0.32 0.39
p04  0.62 0.56 0.55 1.00 0.70 0.27 0.24 0.45 0.24 0.31
p05  0.54 0.51 0.52 0.70 1.00 0.20 0.18 0.35 0.13 0.24
p06  0.23 0.24 0.35 0.27 0.20 1.00 0.58 0.51 0.58 0.46
p07  0.27 0.27 0.31 0.24 0.18 0.58 1.00 0.56 0.68 0.51
p08  0.40 0.44 0.42 0.45 0.35 0.51 0.56 1.00 0.49 0.48
p09  0.24 0.27 0.32 0.24 0.13 0.58 0.68 0.49 1.00 0.54
p10  0.28 0.40 0.39 0.31 0.24 0.46 0.51 0.48 0.54 1.00
```

The matrix variable class contains rectangular data, just like a data frame, but all entries have the same variable type (number, character, logical). You access the data in a matrix by its index numbers, like a data frame: c_matrix[row, col]. The matrix variable class acts like a matrix in linear algebra.

This analysis shows that all correlations are positive, which is a first confirmation of the reliability of the PII data. If the questions did not all relate to an underlying latent variable, then the correlation matrix would be less uniform. Those items will show a negative correlation if the polarity is not corrected.

A scatter plot visualises correlations on the *x*- and *y*-axes. The geom_point() geometry in the *ggplot2* package creates scatter plots. Visualising the data from the survey in this way is problematic because we only have responses between 1 and 7, and many points will be plotted on top of each other, so-called overplotting. One of the solutions to this problem is to add a jitter to the data. Jitter is a small random amount of variation applied to each data point. The *ggplot2* package uses the jitter geometry (geom_jitter()) to implement this technique (Figure 8.4). The width and height variables determine the spread of the points. The alpha parameter sets the opacity of the points, with zero being totally transparent. This way, points that are on top of each other are darker.

```
library(ggplot2)
ggplot(pii, aes(p01, p02)) +
  geom_jitter(width = .5, height = .5, alpha = .5, size = 2) +
  labs(title = "Scatterplot of items p01 and p02",
       subtitle = paste("Correlation:",
                        round(cor(pii$p01, pii$p02), 2))) +
  theme_bw(base_size = 10)
```

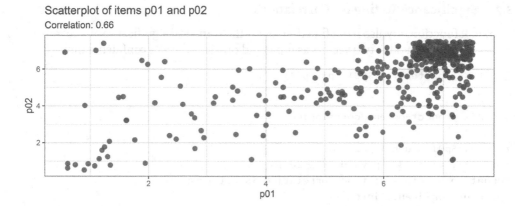

FIGURE 8.4 Scatterplot of items p01 and p02.

Several specialised R packages provide functions to visualise a correlation matrix, such as the *ggcorrplot* package (Kassambara, 2022). This package is not the only library that can create such plots. The power of R is that there are many alternatives, each with its own advantages and disadvantages. The main advantage of using packages that implement the Grammar of Graphics approach is that you can combine layers from different packages.

```
library(ggcorrplot)
ggcorrplot(c_matrix, outline.col = "white") +
  labs(title = "Personal Inventory Index",
       subtitle = "Correlation Matrix")
```

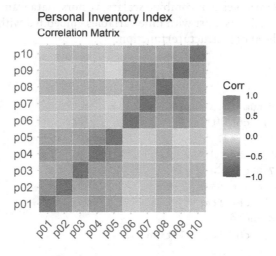

FIGURE 8.5 Correlation matrix for PII.

Figure 8.5 shows that the first five items correlate more strongly with each other than with the other five items, and vice versa. Item p08 seems to relate to all other items. This result seems to indicate that our results confirm the measurement model for the PII, but calculating the correlation is only the first step in assessing the reliability of a survey. The *ggcorrplot* help shows how to create different versions of this visualisation.

8.3.2   Significance Testing for Correlations

The basic R functionality also has a function to test the statistical significance of a correlation. The cor.test() function takes two vectors as input and provides the 95% confidence interval.

```
(c_test <- cor.test(pii$p01, pii$p02))
```

```
Pearson's product-moment correlation

data:  pii$p01 and pii$p02
t = 18.273, df = 437, p-value < 2.2e-16
alternative hypothesis: true correlation is not equal to 0
95 percent confidence interval:
 0.6016030 0.7081116
sample estimates:
      cor
0.6581375
```

The output provides a wealth of statistical information about this correlation. The *t* and *df* values relate to the significance statistics, which uses a *t*-distribution. The low *p*-value tells us that the relationship between these two variables is coincidental is very small. In social science, a value of less than 0.05 is statistically significant in most research. However, you must be careful in interpreting this outcome because a correlation is only a starting point for further analysis. A strong correlation is an invitation to undertake further research. The old adage "correlation is not causation" certainly is valid in this case.

The output of this analysis is a list, which is another type of R variable. You have already seen scalars, vectors, data frames, and matrices. A list is the most flexible type of data and is often used to store the results of an analysis. Most complex analytical functions in R provide output as a list. A list is a collection of R variables that combine scalars, vectors, data frames, and matrices in one structure. The c_test variable is a list with nine variables embedded within it. You can view the structure of a list with the str() (structure) function.

```
str(c_test)
```

```
List of 9
 $ statistic  : Named num 18.3
  ..- attr(*, "names")= chr "t"
 $ parameter  : Named int 437
  ..- attr(*, "names")= chr "df"
 $ p.value    : num 7.86e-56
 $ estimate   : Named num 0.658
  ..- attr(*, "names")= chr "cor"
 $ null.value : Named num 0
  ..- attr(*, "names")= chr "correlation"
 $ alternative: chr "two.sided"
 $ method     : chr "Pearson's product-moment correlation"
 $ data.name  : chr "pii$p01 and pii$p02"
 $ conf.int   : num [1:2] 0.602 0.708
  ..- attr(*, "conf.level")= num 0.95
 - attr(*, "class")= chr "htest"
```

You can access the subsets of a list with the dollar sign indicator. To, for example, display only the *p*-value, use c_test$p.value.

```
c_test$p.value
```

```
[1] 7.861037e-56
```

The `ggcorrplot()` function has the `insig = "blank"` option to plot a correlation matrix, leaving all statistically insignificant correlations blank. In this case study, all correlations in the matrix are significant at the $p < 0.05$ level.

```
ggcorrplot(c_matrix, insig = "blank", show.diag = FALSE)
```

The strong correlation between the survey items does not mean these responses *cause* each other. Instead, the correlation indicates that there might be an underlying phenomenon that causes them to correlate. We hypothesise that this cause is the psychological construct of consumer involvement, which we set out to measure.

8.3.3 Measuring Survey Reliability with Cronbach's Alpha

Strong correlations between survey items indicate high reliability and validity, but it is insufficient evidence. The most commonly used method to assess the reliability of a survey instrument is Cronbach's Alpha or coefficient alpha. This coefficient measures how closely related a set of items are as a group. This coefficient is between zero and one, with one implying total consistency. An alpha value larger than 0.7 is generally considered acceptable. The higher the coefficient, the higher the reliability of the survey (Cronbach & Meehl, 1955). Surveys published in scholarly literature almost always mention Cronbach's Alpha as a measure of reliability (Bruner, 2012; Bearden, 2011).

This coefficient uses the covariance between the survey items. Covariance is the degree to which the deviation of a variable from the mean relates to the deviation of another variable from its mean. Covariance is similar to correlation. However, correlation looks at how two variables interact with each other in strength and direction instead of their shared variance. Correlation values range from plus one to minus one, while covariance can take any value (Equation 8.1).

$$cov(x, y) = \frac{\sum (x_i - \bar{x})(y_i - \bar{y})}{n - 1} \tag{8.1}$$

Covariance and correlations are closely related. The correlation between two variables x and y is their covariance divided by the product of their standard deviations (Equation 8.2).

$$cor(x, y) = \frac{cov(x, y)}{s_x s_y} \tag{8.2}$$

The code below calculates the covariance and correlation between two items from the survey using these formulas and the relevant R functions as a comparison. The `with()` function is a convenient method to not have to repeat the data frame name every time you use a variable. For example: `with(pii, p01 - mean(p01))` gives the same result as `pii$p01 - mean(pii$p01)`.

```
with(pii, sum((p01 - mean(p01)) * (p02 - mean(p02)))) / (nrow(pii) - 1)
```

```
[1] 1.674931
```

```
cov(pii$p01, pii$p02)
```

```
[1] 1.674931
```

```
cov(pii$p01, pii$p02) / (sd(pii$p01) * sd(pii$p02))
```

```
[1] 0.6581375
```

```
cor(pii$p01, pii$p02)
```

```
[1] 0.6581375
```

The cov() function works in the same way as the correlation function. To create a covariance matrix use:

```
round(cov(pii[, -1]), 2)
```

```
      p01  p02  p03  p04  p05  p06  p07  p08  p09  p10
p01  2.52 1.67 1.10 1.44 1.27 0.66 0.80 1.18 0.69 0.75
p02  1.67 2.57 1.20 1.32 1.20 0.70 0.80 1.31 0.78 1.08
p03  1.10 1.20 2.34 1.24 1.17 0.97 0.89 1.18 0.89 1.01
p04  1.44 1.32 1.24 2.16 1.50 0.72 0.65 1.21 0.63 0.77
p05  1.27 1.20 1.17 1.50 2.17 0.54 0.50 0.95 0.34 0.59
p06  0.66 0.70 0.97 0.72 0.54 3.25 1.94 1.70 1.88 1.40
p07  0.80 0.80 0.89 0.65 0.50 1.94 3.44 1.93 2.28 1.60
p08  1.18 1.31 1.18 1.21 0.95 1.70 1.93 3.39 1.63 1.52
p09  0.69 0.78 0.89 0.63 0.34 1.88 2.28 1.63 3.29 1.66
p10  0.75 1.08 1.01 0.77 0.59 1.40 1.60 1.52 1.66 2.89
```

The formula for Cronbach's Alpha uses the number of survey items k, the average covariance between these items $\bar{c}$ and the average variance $\bar{v}$ (Equation 8.3).

$$\alpha = \frac{k\bar{c}}{\bar{v} + (k-1)\bar{c}} \tag{8.3}$$

The variance of an item is the same as the covariance with itself. In other words, it is the same as the diagonal of the covariance matrix. The diag() function extracts the diagonal from a matrix. To determine the mean covariance, we only need the lower or upper triangle of the covariance matrix. The result of the lower.tri() and upper.tri() functions is a matrix with the same size as the input with TRUE / FALSE indicators for either the diagonal or one of the triangles.

```
pii_cov <- cov(pii[, -1])
k <- ncol(pii_cov)
v <- mean(diag(pii_cov))
c <- mean(pii_cov[lower.tri(pii_cov)])
alpha <- (k * c) / (v + (k - 1) * c)
alpha
```

```
[1] 0.8726885
```

We see that coefficient alpha for the PII survey is greater than 0.7 and thus sufficient to consider this survey instrument reliable.

Manually calculating Cronbach's Alpha takes a few steps and is only included to demonstrate how R can manipulate matrices. The *psych* package provides advanced functionality to analyse psychometric data (Revelle, 2022), including a function to calculate Cronbach's Alpha. This function provides a very detailed output beyond this book's scope. The part we need for this case study is the raw alpha, using the covariance matrix tucked away inside a list.

```
alpha <- psych::alpha(pii[, -1])
alpha$total$raw_alpha
```

```
[1] 0.8726885
```

8.4 Survey Validity

The objective of a good survey is to find a causal relationship between the mental state, the latent variable and the survey questions (Figure 8.1). The validity of the survey is a measure of the strength of this relationship. Various methods exist that address this problem; some of these are (DeVellis, 2011):

- *Content validity*: A judgement about whether the survey instrument captures all the relevant components of the latent variable.

- *Construct validity*: Does the construct (the way the latent variable is measured) describe the latent variable?

Looking at the consumer involvement survey items, they all seem to relate to what we intuitively understand this construct. A formal content validity assessment would involve a panel of experts to review the survey items. Construct validity is a more quantitative approach that assesses whether a single underlying cause relates to all items in the survey. Exploratory Factor Analysis is a statistical technique that looks for such patterns in a data matrix.

8.4.1 Exploratory Factor Analysis

The customer's state of mind cannot be directly observed, and we can only infer these variables by analysing manifest (observable) variables. The manifest variables in the case of a survey are the answers that respondents provide to the individual questions. Exploratory factor analysis uncovers structure in the survey data by analysing the correlation matrix. The results of factor analysis offer insight into the relationship between the observable variables (the survey responses) and the latent variables that reside in the mind of the respondents.

This technique determines to which extent items in a survey instrument relate to each other. The variables that relate the strongest to each other form a so-called factor, which means that we reduce the complexity of the survey, in this case, ten items, to one or more factors. Each observed variable is potentially a measure of each of the hypothesised factors through a linear relationship. The results determine which items are part of the same factor. For example, the theoretical model for the involvement inventory defines that five items relate to the affective dimension and the other five to the cognitive dimension (Zaichkowsky, 1994).

The relationship between factors and the items is never perfect. The extent to which a factor describes the underlying variables is expressed in the amount of variance the factor explains. The mathematical principles underpinning factor analysis are much like linear regression, discussed in Chapter 9. This chapter only provides a glimpse of this technique, as there are a lot of statistical subtleties with using this technique that are outside the scope of this book.

The first decision is to determine the number of factors to extract. Some statistical methods are available, but in this case, we follow the theoretical structure designed by the authors of this scale. The Personal Involvement Inventory consists of two dimensions, which means we summarise the ten measured variables in two latent variables.

Factors are extracted from correlation matrices by transforming such matrices by eigenvectors. An eigenvector expresses how much of the variance the model explains. A perfect model would explain all variance where the calculated values perfectly match the observed values. But all measurements have errors, and our model is rarely a perfect representation of the measurements. The mathematical model for factor analysis looks for the best combination of linear factor loadings that explains most of the variance. In the mathematical model, p denotes the number of variables $(X_1, X_2, \ldots, X_p)$, which are the ten survey items in our case study. The letter m denotes the number of underlying factors $(F_1, F_2, \ldots, F_m)$, which in this case study is two. Factor analysis assumes that there are m underlying factors, and each observed variable is a linear function of these factors together with a residual variate. Factor analysis searches for the model that best explains all the variance in the data. Each variable is thus modelled by (Yong & Pearce, 2013):

$$X_j = a_{j1}F_1 + a_{j2}F_2 + \ldots + a_{jm}F_m + \epsilon_j \qquad (8.4)$$

Where X_j expresses the observed variables and $a_{j1}, a_{j2}, \ldots, a_{jm}$ the loadings (regression coefficients) for each of the factors. The outcome of factor analysis is thus a regression matrix with coefficients that relate each factor to each observable variable. Any factor loading smaller than 0.3 is usually ignored, and the ideal situation, the so-called Simple Model, is that each item only loads on one factor. When we create a simple model, we can calculate the latent variable using the measurable variables and the factor scores as input.

The results of factor analysis can be challenging to interpret. Rotating the results can help to clarify the results. Rotation geometrically pivots the results in m-dimensional space to find the perspective that provides the most information. This rotation is analogous to rotating a cube where you either see three surfaces or only one or two.

8.4.2   Factor Analysis with R

Factor analysis forms part of the base R functionality, but this case study uses the more advanced version in the *psych* package. The `fa()` function takes the data frame as the first parameter, and you need to define the number of factors you like to extract. You also need to specify the required type of rotation, which requires the *GPArotation* package (Bernaards & I.Jennrich, 2005). A range of rotation methods are available. The choice of rotation method depends on the type of data and the purpose of the analysis. The oblimin rotation is the most common rotation method.

The output of the `fa()` function is extensive, and we only need a part of it for this case study. The result is stored in a list that we can investigate. The `loadings` variable inside the list contains the most salient information.

```
library(psych, quietly = TRUE)
library(GPArotation)
pii_fa <- fa(pii[, -1], nfactors = 2, rotate = "oblimin")
pii_fa$loadings
```

```
Loadings:
      MR1    MR2
p01  0.746
p02  0.698
p03  0.569  0.199
p04  0.845
p05  0.810 -0.103
p06         0.704
p07         0.834
p08  0.300  0.551
p09         0.841
p10  0.150  0.594

                MR1    MR2
SS loadings     2.858  2.612
Proportion Var  0.286  0.261
Cumulative Var  0.286  0.547
```

The output shows the loadings (ranging between -1 and 1) for each survey item for each extracted factor MR1 and MR2. These numbers express the contribution of each item to the latent variables. Values lower than 0.1 are removed for clarity. We are looking for loadings greater than 0.3 to assign each variable to a factor. This table suggests that items p01 to p05 load highest on the first factor and the remainder of the items on the second factor. Item p08 seems to load on both factors, but it is only a marginal case.

Should any of the items not significantly load on any of the factors, then these items are removed, and the analysis is repeated. When developing a scale, researchers often develop more items than strictly needed and then use factor analysis to select the most effective survey questions.

The table beneath the loadings shows the proportion of variance explained by each factor. The Cumulative Var row displays the cumulative proportion of explained variance, ranging from zero to one. The Proportion Var row shows the proportion of variance that each factor explains, and the row SS loadings the sums of the squared loadings. A factor is worth keeping if the SS loading is greater than 1 (Kaiser Criterion), which is easily the case for this analysis.

8.4.3 Visualising Factor Analysis

The concept of factor rotation is best explained by visualising it. The code example below undertakes two factor analyses, one without rotation and one with the oblimin model. The graphs in Figure 8.6 show the two factors on the *x*- and *y*-axes, and each observable variable is plotted with their respective loading for each factor. The rotation moves the factors closer to axes so that the factor loadings are maximised on one factor and minimised on the other, reducing cross-loading.

```
par(mfcol = c(1, 2))
fa.none <- factanal(pii[, -1], factors = 2, rotation = "none")
fa.oblimin <- factanal(pii[, -1], factors = 2, rotation = "oblimin")

plot(fa.none$loadings[,1],
     fa.none$loadings[,2],
     pch = 19, col - "darkgrey",
     xlab = "Factor 1", ylab = "Factor 2",
     ylim = c(-1,1), xlim = c(-1,1),
     main = "No rotation")
abline(h = 0, col = "grey")
abline(v = 0, col = "grey")

plot(fa.oblimin$loadings[,1],
     fa.oblimin$loadings[,2],
     pch = 19, col = "darkgrey",
     xlab = "Factor 1", ylab = "Factor 2",
     ylim = c(-1,1), xlim = c(-1,1),
     main = "Oblimin rotation")
abline(h = 0, col = "grey")
abline(v = 0, col = "grey")

text(fa.oblimin$loadings[,1] + 0.04, fa.oblimin$loadings[,2] + 0.04,
     colnames(pii[, -1]), cex = 0.7)
```

The code below shows how the *psych* package visualises the measurement model for the PII. The fm = "ml" parameter ensures that this function uses the same factor extraction method as the function in the base package. In addition, the *psych* version can implement other extraction methods required in specialised cases.

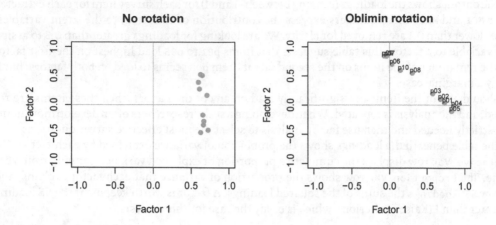

FIGURE 8.6 Unrotated and rotated factors for Personal Involvement Index.

Figure 8.7 is a typical representation of a measurement model for a psychological construct. The squares represent the observable variables, which are in this case the survey items. The ML1 and ML2 variables are the two dimensions of the scale and are denoted with ovals. The direction of the arrows shows that the latent variables are hypothesised to cause the survey responses. The two dimensions themselves are also correlated, which confirms the theoretical model for PII. This correlation is strengthened by the rotation method used to optimise the factors.

```
library(psych)
pii_fa <- fa(pii[, -1], nfactors = 2, rotate = "oblimin", fm = "ml")
fa.diagram(pii_fa, main = NULL)
```

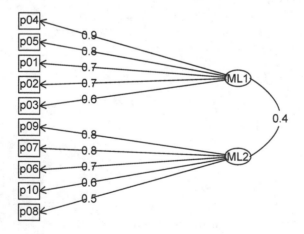

FIGURE 8.7 Visualising the measurement model.

Following this analysis, we can conclude that the ten PII items measure an underlying attitude towards water. We can conclude from the factor solution that the PII is a single latent construct that describes how consumers feel about tap water from a cognitive and affective perspective.

8.5 Interpreting Consumer Involvement

Now that we have confirmed that we can use a two-factor model, we can reduce the ten measured variables to two or even one latent variables and use this in further analysis. Factor analysis provides a linear model that describes the relationship between the measured and latent variables.

The final task is to determine what these factors represent. Table 8.1 lists the survey participants' actual questions. The first five questions relate to more rational aspects of consumer involvement, while the last five are more emotive. The analysis has thus confirmed a significant difference between the two dimensions of consumer involvement. The beauty of factor analysis is that the linguistic difference between the questions becomes apparent in the numeric analysis of the answers.

```
library(tidyr)
pii_scores <- pii %>%
  mutate(cognitive = p01 + p02 + p03 + p04 + p05,
         affective = p06 + p07 + p08 + p09 + p10) %>%
  select(customer_id, cognitive, affective)

pivot_longer(pii_scores, cols = -customer_id) %>%
  ggplot(aes(value)) +
  geom_histogram(fill = "dodgerblue 4") +
  facet_wrap(~name) +
  theme_minimal(base_size = 12) +
  labs(title = "Consumer Involvement with Tap Water",
       subtitle = "Personal Involvement Index")
```

The factor model only assigns a measurable variable to a latent variable. Therefore, we sum the response scores for each factor to calculate the involvement scores. In other words, after the analysis, the factor loadings are assumed to be 1. The code below calculates each respondent's two dimensions of consumer involvement and visualises the result in Figure 8.8.

These histograms show that the level of affective involvement is significantly lower than the level of cognitive involvement. This outcome is not a surprise as it seems intuitively correct that tap water consumers intellectually care about water as an essential ingredient in life. Still, their emotional connection with the service is not as significant.

8.6 Further Study

This chapter introduced the R matrix variable type. R can handle a matrix in the same way as a data frame, but also has linear algebra functions for matrix calculations. The %*% operator performs the dot product of two matrices.

Measuring the customer experience with surveys is a complex craft. The *Questionnaire Design for Social Surveys* course from the University of Maryland goes into great depth on how to best design surveys.

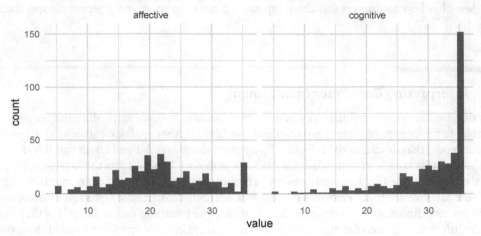

FIGURE 8.8 Personal Involvement Index scores.

Accurate measurement of psychological constructs is a complex topic that goes beyond the scope of this course. Please note that the examples in this chapter do not constitute a thorough analysis of latent constructs. Correlations and cluster analysis are great for exploration. Structural equation modelling is best practice in psychographic analysis. If you are interested in the statistical intricacies of measuring the customer experience, then read *Scale Development: Theory and Applications* by Robert DeVellis (2011).

9

Basic Linear Regression

The previous two chapters demonstrated some basic principles of administering and analysing customer survey data. Chapter 8 shows how to use a correlation matrix to find patterns in survey responses. In that case, we had justified reasons to believe that the correlation was caused by an underlying latent variable. A correlation can also mean that the variables influence each other, which is the domain of regression analysis.

Regression analysis is one of the most common methods to investigate relationships between variables. Understanding linear regression is the first step toward predictive analysis and machine learning. This chapter explores possible linear relationships between the responses in the customer survey and uses these results to explain the theory and practice of building and to assess linear models. The learning objectives for this chapter are:

- Understand the principles of linear regression

- Perform a linear regression of the customer survey data

- Assess the significance of a linear regression

9.1 Principles of Linear Regression

The purpose of a regression model is to predict one variable by measuring one or more other variables through a linear relationship. For example, you might want to predict water consumption based on the forecast temperature or investigate how customer complaints relate to the level of pressure.

The outcome of regression analysis is a function that describes this relationship that predicts unobserved future events. Through regression, the computer learns from existing observations to predict future events.

Researchers manipulate the independent variables in experimental settings to study their effect on the dependent variable. For example, in an urban water context, a standard dependent variable is water consumption, and independent variables are parameters such as garden size, outdoor temperature, and so on (Cominola et al., 2021).

Figure 9.1 shows an example of a linear regression using random numbers. Regression analysis finds the line of best fit through the points x_i and y_i (the solid line). The best fit is the line that minimises error, indicated by the dotted lines. The error ϵ, also called residual, is the absolute difference between the measured value and the predicted value $\hat{y}$ (pronounced y-hat).

$$y_i = \beta_0 + \beta_1 x + \epsilon_i \tag{9.1}$$

$$\hat{y}_i = \beta_0 + \beta_1 x \tag{9.2}$$

The most common method to determine the parameters for this line is Ordinary Least Squares. The *Sum of Squares* (SS_{fit}) of the difference between the predicted (fitted) values $\hat{y}$ and the observed values y_i indicates how well the predicted results fit the observations.

To find the best fit, rotate a line through the mean values line around the centre of the point cloud ($\bar{x}, \bar{y}$) until you find the lowest sum of squares (SS). Figure 9.1 visualises the residuals for the average y value, two rotations and the model with the lowest SS value.

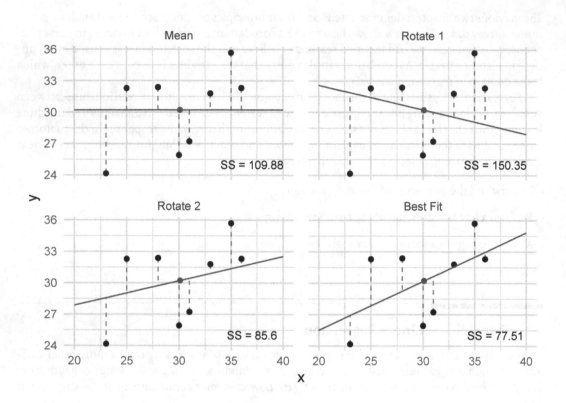

FIGURE 9.1 Ordinary Minimum Least Squares method.

The slope of the line relates to the correlation and the ratio of the standard deviations of the observed variables. Once the slope is known, the intercept is also known. We can determine the linear parameters β_0 and β_1 with the following formulas:

$$\beta_1 = cor(y, x)\frac{s_y}{s_x} \tag{9.3}$$

$$\beta_0 = \bar{y} - \beta_1\bar{x} \tag{9.4}$$

The data in the example in Figure 9.1 is stored in the d variable. We can use basic R functions to calculate the parameters for the regression line

```
set.seed(1066)
 d <- tibble(x = sample(22:38, 8),
             y = runif(1) * x + rnorm(8))
```

```
beta_1 <- cor(d$y, d$x) * sd(d$y) / sd(d$x)
beta_1
```

```
[1] 0.5960773
```

```
beta_0 <- mean(d$y) - beta_1 * mean(d$x)
beta_0
```

```
[1] 3.222573
```

The closer the residuals are to zero, the better the fit and the more reliable any prediction will be. If the measured points would fall perfectly in one line, then the Sum of Squares is zero, and we have a perfect fit. This basic calculation from first principles is only an example. The R language has extensive capabilities to build and assess linear models.

9.2 Basic Linear Regression in R

The R language has a versatile linear modelling function to analyse data and assess the model's reliability. Let's use some of the data from Case Study two. Customers were asked to indicate whether they struggle to pay their water bills when they fall due. This question used a seven-point Likert scale from "Strongly Disagree" to "Strongly Agree", which we can code 1–7. The second question asked customers to indicate the frequency at which they contact their utility for support, also using a seven-point Likert scale:

1. Never

2. Less than once a month

3. Once a month

4. 2–3 times a month

5. Once a week

6. 2–3 times a week

7. Daily

Perhaps we can use this data to predict whether a customer might experience hardship by looking at their actual contact frequency. The hypothesis is that customers who suffer from financial hardship will contact the utility more frequently than those who easily pay their bills. To analyse this hypothesised relationship, we use the cleaned survey data prepared in Chapter 7, extract the data we need, and remove missing observations.

```
source("scripts/07-customer_clean.R")
```

```
cont_hard <- select(customers, contact, hardship)
cont_hard <- cont_hard[complete.cases(cont_hard), ]
```

The first step in our workflow is to visualise the data. Visualising the relationships between variables is essential, as shown in Figures 4.6 and 6.2. Because Likert-scale data only contains the integers 1–7, we have an overplotting issue. In the previous chapter, we solved it with random jitter (Figure 8.4). Another method to prevent plotting points on top of each other is to count the frequency of each combination of hardship and contact and relate the point size to the frequency

(Figure 9.2). Finally, the `scale_size()` function scales the size of the point geometry by sizing the point area to the frequency.

The smoothing layer draws a linear regression model. We need to tell it to use the original data and not the frequency table generated in the first line. This graph is an example of how layers in *ggplot2* can use different data sets by using `data` and `aes()` in the geometry function. The regression line from the smoothing geometry suggests a linear relationship between the two variables. The grey band behind the regression line is the 95th percentile confidence interval, which is discussed in more detail below.

```
count(cont_hard, hardship, contact) %>%
  ggplot() +
    geom_point(aes(contact, hardship, size = n), col = "darkgrey") +
    scale_size(guide = "none", range = c(0, 20)) +
    geom_smooth(data = customers, aes(contact, hardship), method = "lm") +
    labs(title = "Gormsey Customer Survey",
         x = "Contact frequency", y = "Financial hardship") +
    theme_light(base_size = 10)
```

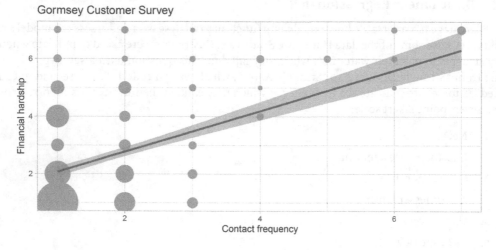

FIGURE 9.2 Linear regression between financial hardship and contact frequency.

9.3 The Linear Model Function

The `lm()` function performs linear regression models. The linear modelling function in the R base library provides detailed information about the analysis, which is stored in a list.

```
hc_model <- lm(hardship ~ contact, data = cont_hard)
```

The `lm()` function uses the formula notation common in R data models. You need to read this syntax as "the linear model (`lm`) of `hardship` predicted by (`~`) `contact` from the `customers` data frame". You can find the tilde (`~`) symbol near the escape key. This notation is the most simple format for a regression model. Additionally, the formula notation in R can describe complex models with transformed variables, multiple regressions, and other interactions between variables.

Printing the results to the console shows the main results. The print function recognises that hc_model is a linear model and formats the output accordingly, displaying the intercept value β_0 and the regression coefficient β_1.

```
hc_model
```

```
Call:
lm(formula = hardship ~ contact, data = cont_hard)

Coefficients:
(Intercept)        contact
     1.3654         0.7052
```

This formula suggests that customers who contact the utility more often experience a higher level of hardship as the slope coefficient is positive. While this conclusion is intuitively correct, are we mathematically justified in accepting this relationship?

9.4 Assessing Linear Relationship Models

Linear model list variables in R contain a wealth of diagnostic information about the analysis. We can delve deeper into the analysis with the summary() function. This function recognises the linear model list and formats the output accordingly.

```
summary(hc_model)
```

```
Call:
lm(formula = hardship ~ contact, data = cont_hard)

Residuals:
    Min     1Q  Median     3Q     Max
-5.3016 -1.0706 -0.7757  1.2243  4.9294

Coefficients:
            Estimate Std. Error t value Pr(>|t|)
(Intercept)  1.36537    0.13102   10.42   <2e-16 ***
contact      0.70518    0.06147   11.47   <2e-16 ***
---
Signif. codes:  0 '***' 0.001 '**' 0.01 '*' 0.05 '.' 0.1 ' ' 1

Residual standard error: 1.623 on 433 degrees of freedom
Multiple R-squared:  0.2331,Adjusted R-squared:  0.2313
F-statistic: 131.6 on 1 and 433 DF,  p-value: < 2.2e-16
```

The first part of the summary shows the function call that created it. This information is helpful if you are running more than one linear model, so there is no confusion about how the results were derived.

9.4.1   Residuals

After the function call, the first bit of information summarises the residuals. You can calculate these residuals and compare them with the model output differently. The `predict()` function uses the output of a model to estimate the predicted values. The code below calculates the residuals as a demonstration, but they are also stored in the `hc_model$residuals` variable in the model or are available through the `residuals()` function.

```
cont_hard %>%
  filter(!is.na(hardship) & !is.na(contact)) %>%
  mutate(prediction = predict(hc_model),
         residual = residuals(hc_model),
         res_calc = hardship - prediction,
         res_lm = hc_model$residuals)
```

```
# A tibble: 435 × 6
   contact hardship prediction residual res_calc  res_lm
     <dbl>    <dbl>      <dbl>    <dbl>    <dbl>   <dbl>
 1       2        1       2.78    -1.78    -1.78   -1.78
 2       7        7       6.30     0.698    0.698   0.698
 3       2        5       2.78     2.22     2.22    2.22
 4       1        1       2.07    -1.07    -1.07   -1.07
 5       1        2       2.07    -0.0706  -0.0706 -0.0706
 6       1        1       2.07    -1.07    -1.07   -1.07
 7       1        4       2.07     1.93     1.93    1.93
 8       1        5       2.07     2.93     2.93    2.93
 9       4        4       4.19    -0.186   -0.186  -0.186
10       1        2       2.07    -0.0706  -0.0706 -0.0706
# ... with 425 more rows
# i Use `print(n = ...)` to see more rows
```

The output of the linear model in R shows the result of the `summary()` function over the residuals (Section 4.3). One of the assumptions for hypothesis testing is that errors follow a normal distribution, and so should the residuals. Therefore, the median should be zero because the mean of the residuals is zero by definition, and in a normal distribution, the mean and median are equal. Also, the first and third quartile (1Q and 3Q) and the minimum and maximum should mirror each other under a normal distribution.

```
summary(hc_model$residuals)
```

```
   Min. 1st Qu.  Median    Mean 3rd Qu.    Max.
-5.3016 -1.0706 -0.7757  0.0000  1.2243  4.9294
```

We can also visualise the residuals as a histogram with the `hist()` function, which is quicker than the more formal `ggplot2` version, requiring you to create a data frame first. The built-in plotting functions are ideal for quickly exploring data (Figure 9.4).

```
hist(residuals(hc_model), breaks = 20)
```

Violating these symmetries either indicates that the model does not fit the observation or that there are outliers that skew the data. Looking at numbers or a histogram is an informal method to test a distribution for normality. You can test a vector for normality with the Shapiro-Wilk normality test. The Shapiro–Wilk test tests the null-hypothesis that a sample is taken from a normally distributed population. The test parameter W expresses the ratio of the variance of the sample and a simulated normal distribution.

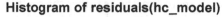

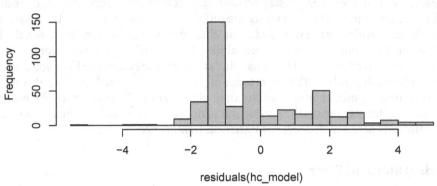

FIGURE 9.3 Histogram of residuals.

```
shapiro.test(hc_model$residuals)


	Shapiro-Wilk normality test

data:  hc_model$residuals
W = 0.90348, p-value = 5.552e-16
```

If the p value is less than the chosen alpha level (often 0.05), then the null-hypothesis is rejected, and there is evidence that the data tested are not normally distributed. Please note that statistical significance testing is unreliable with massive samples. However, the results of this test indicate that we can assume the residuals to be normally distributed because the p value is close to zero, which means that the likelihood that this sample of residuals was *not* derived from a normally-distributed population is close to zero.

9.4.2 Coefficients

The next part of the summary provides information about the estimated regression coefficients, their standard errors, t statistics, and p values. The estimated coefficients express a formula we saw in the previous section. The intercept is essentially the predicted value when the regressor is zero. You can also extract the coefficients from the model with the coef() function, which extracts the coefficient variable from the results list into a vector. Alternatively, use hc_model$coefficients to achieve the same result.

```
coef(hc_model)

(Intercept)     contact
  1.3653727   0.7051781
```

The Standard (Std.) Error in the detailed summary is the residual standard error (see below) divided by the square root of the sum of the square of that particular variable. This number expresses the uncertainty in the estimated coefficient. You can use the standard error to construct confidence intervals.

The null-hypothesis in linear regression is that the beta coefficients associated with the variables are equal to zero. The alternate hypothesis is that the coefficients are not equal to zero. If the null-hypothesis is refuted, then there exists a relationship between the independent and the dependent variables. The linear model function tests the null-hypothesis with a t-test. The t value is the regression coefficient divided by the standard error. Thus, the greater the standard error, the lower the t value. The `Pr(>|t|)` column shows the probability that the null-hypothesis is confirmed with the relevant degrees of freedom. The closer this value is to zero, the better. The limit you are willing to accept as significant depends on the importance of the conclusion you draw from this number. The stars (`***`) behind the probabilities indicate their significance. The more stars, the higher the level of significance. One star means the p-value is less than 0.05. This analysis thus shows that the regression coefficients have a high degree of statistical significance.

9.4.3 Residual Standard Error

The residual standard error measures how well the model fits the data. This value is the square root of the average sum of squares, divided by the degrees of freedom (Equation 9.5). The closer this value is to zero, the better the model fits the data.

```
coef(hc_model)
```

$$rse = \sqrt{\frac{\sum (y - \hat{y})^2}{df}} \tag{9.5}$$

The degree of freedom df relates to the number of observations and regressors. In a model with only two observations, there are zero degrees of freedom because there is only one line that you can draw through these lines. The degree of freedom for regression with one parameter is $df = n - 2$. In our case study, we have 435 rows of data and one regressor (`contact`), so $df = 433$.

```
n <- nrow(cont_hard)
k <- length(hc_model$coefficients) - 1
ss_fit <- sum(hc_model$residuals^2)
df <- n - (k + 1)
rse <- sqrt(ss_fit / df)
rse
```

```
[1] 1.622513
```

This standard error is somewhat high, as it spans a fair part of the measurement domain between 1 and 7.

9.4.4 R-Squared

The R-Squared value expresses the proportion of the variance in the observations that the model explains. The model explains about 23% of the variance in the case study. This result means that the model explains about a quarter of the variance in the measured data. This low value does not imply a relationship between these variables. More than likely, other confounding variables, such as average debt or time to pay a bill, predict hardship status. However, caution is needed when adding additional regressors as this can increase R^2, even if the additional information has no relationship with the dependent variable. You can extract the R^2 value from the summary list by calling the `r.squared` variable.

```
summary(hc_model)$r.squared
```

```
[1] 0.233078
```

For two variables, the R^2 value is the square of the correlation, but this simple approach does not hold for more complex regression models. To calculate R^2 from first principles, you need the variance of the observations, which is the average of the Sum of Squares (SS) from the sample mean ($\bar{y}$) and the variance of the fitted values ($\hat{y}$). The proportion of variance explained by the linear model is the proportion of the difference between the variance of the fitted model and the variance from the mean, divided by the variance from the mean.

$$SS_{fit} = \sum_{i=1}^{n}(y - \hat{y})^2 \tag{9.6}$$

$$SS_{mean} = \sum_{i=1}^{n}(y - \bar{y})^2 \tag{9.7}$$

$$R^2 = \frac{SS_{mean} - SS_{fit}}{SS_{mean}} \tag{9.8}$$

If the residuals of the fitted values are zero (all observations are in a straight line), then the value for R^2 is one. If there is no relationship between the variables, then $R^2 = 0$.

```
ss_fit <- sum(hc_model$residuals^2)

ss_mean <- sum((cont_hard$hardship - mean(cont_hard$hardship))^2)

(ss_mean - ss_fit) / ss_mean
```

```
[1] 0.233078
```

Adjusted R^2 normalises this statistic by considering the number of samples and explanatory variables (degrees of freedom). This adjusted R^2 should be used in reporting and analysis.

```
1 - (ss_fit / ss_mean) * (n - 1) / df
```

```
[1] 0.2313068
```

The R^2 value should be interpreted with care. It can be manipulated by reducing the degrees of freedom by increasing the sample size or adding spurious predictors.

9.4.5 F-statistic

The F-statistic expresses the ratio between the amount of variance from the mean and the amount of variance, corrected for the degrees of freedom. The F-test is a global test of fit that checks if at least one predictor is significantly different from zero. If the p-value is greater than the acceptable limit α, then the model does not predict anything. The F-statistic allows you to test the null-hypothesis that the model describes random data.

$$F = \frac{var_{mean}}{var_{fit}} df \tag{9.9}$$

```
(ss_mean - ss_fit) / (ss_fit) * df
```

```
[1] 131.5946
```

The linear model for a set of randomly-generated variables results in a meagre value for F. However, when creating hundreds of linear models with random data, high values for F are very rare, so the higher the F-statistic, the less likely the results will be caused by mere chance.

9.5   Graphical Assessment

Reviewing residuals graphically provides more information about the linear model. When the `plot()` function processes a linear model, it will display four graphs (Figure 9.4). When plotting them in the console, you need to press enter to view the next graph. You can prevent doing so by splitting the plot screen into a two-by-two grid. The first line of code splits the plot screen into two rows and two columns. Note that this parameter does not work when you use *ggplot2*. The `pch`, `col`, and `cex` options change the shape, colour, and size of the data points.

```
par(mfrow = c(2, 2))
plot(hc_model, pch = 19, col = "grey", cex = .5)
```

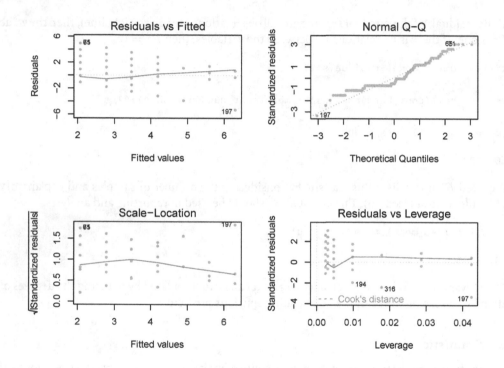

FIGURE 9.4 Analysis of the linear model residuals.

The four plots provide the following information:

1. Residuals versus fitted
2. Normal Q-Q
3. Scale-Location
4. Leverage

9.5.1   Residuals versus Fitted Plot

The first plot visualises whether the residuals exhibit any pattern. If the residuals are genuinely random, as they should be, then this plot does not show any pattern. However, the plot in this case study seems to suggest a pattern instead of randomness caused by overplotting. The red line is a LOWESS regression (Locally Weighted Scatterplot Smoothing) of the residuals. Basically, it smooths the points to look for patterns in the residuals. In a good fitting model, this line is close to the zero line as the residuals are normally distributed, which is the case in this example.

9.5.2 Normal Q-Q Plot

The second plot tests the residuals for normality. In a Quantile-Quantile (Q-Q) plot, the dotted straight line shows the theoretical quantiles of a normal distribution. Standardised residuals are the raw residuals $(y_i - \hat{y})$ divided by their standard deviation. In practice, any standardised residual greater than three are outliers (see Chapter 12). The graph in Figure 9.4 shows that the residuals are close to a normal distribution, as confirmed analytically in Section 9.4.1. The numbered points are considered outliers in the data.

If the residuals do not follow a normal distribution, you could try transforming your data. You could, for example, convert one or two variables to a log or square root scale and recast the model. More formal systems, such as Tukey's Ladder, are available to enforce the condition of normality in your data (Tukey, 1977). Table 9.1 summarises the most common transformation following the ladder.

TABLE 9.1 Tukey's ladder for transforming variables.

Power	Name	Transformation	Inverse
2	Square	x^2	sqrt(x)
1	No transformation	x	x
½	Square root	sqrt(x)	x^2
0	Logarithm	log(x)	exp(x)
-½	Reciprocal root	-1 / sqrt(x)	1/x^2
-1	Reciprocal	-1 / x	-1/x
-2	Reciprocal square	-1/x^2	sqrt(-1 / x)

When transforming a variable in a prediction you will need to use the inverse operation on the predicted values to scale the data back to the original units.

9.5.3 Scale-Location Plot

The third plot displays the fitted values along the x-axis and the square root of the standardised residuals on the y-axis. This plot should also show a random pattern, and the red line should run horizontally, just like the first graph. There should be no recognisable pattern in the plot of the fitted values. When a pattern is apparent, you might have a problem with heteroscedasticity, a situation where the residuals are not random.

9.5.4 Residuals versus Leverage Plot

The last plot (Cook's distance) tells us which points have the most significant influence on the regression (leverage points). In this plot, the pattern is irrelevant, but we need to look for values outside an acceptable Cook's Distance, indicated by points outside the dotted lines. Three observations are singled out and marked as outliers but not outside acceptable limits. The best way to deal with any high leverage points is to remove them from the analysis.

The lm() function has an option to remove outliers from the model. The code below removes three points identified as having significant leverage. Building a model is an iterative process where the analyst resides in the data vortex, as shown in the data science workflow in Figure 6.1. You might have to develop several models until you reach a point where the outcome meets the requirements of the problem statement.

```
hc_model2 <- lm(hardship ~ contact, data = cont_hard,
                subset = c(-194, -197, -316))
```

9.6 Polynomial Regression

Developing more complex models follows the same process as creating a simple model. The main difference is in the model definition. The linear regression function can also model polynomial equations using the identity function I().

The formula for channel flows in Section 2.4 describes the relationship between the measured height and flow in a rectangular channel. This formula includes a discharge coefficient C_d, which we could derive empirically. Let's assume we did some measurements in a laboratory where we measure the volume with a flow meter and measure the height over the weir. We can then use linear regression to derive the discharge coefficient.

The code below generates flow measurements with some random error. The set.seed() function fixes the random number generator so that every time the code runs it generates the same pseudo-random numbers. This method is often used in stochastic analysis to ensure the analysis is reproducible.

The lm() function predicts the curve that best fits the data. Note how the model now includes I(h^(2/3)), which instructs R to execute what is between parentheses instead of interpreting it as part of the formula. In this case, the function call regresses the observations to $h^{2/3}$. The -1 in the function call sets the y-axis intercept value to zero. Figure 9.5 shows the observations, the theoretical values (dotted line) and the model (solid line).

```
set.seed(1969)
g <- 9.81
b <- 0.5
Cd <- 0.62

h <- seq(from = 0, to = 0.2, by = 0.01)
q_observed <- 2/3 * sqrt(2 * g) * b *
  rnorm(length(h), mean = Cd, sd = .1) * h^(2/3)
q_theory <- 2/3 * sqrt(2 * g) * b * Cd * h^(2/3)

model <- lm(q_observed ~ I(h^(2/3)) - 1)

par(mar = c(4, 4, 1, 1), mfrow = c(1, 1))
plot(h, q_observed, pch = 19)
lines(h, q_theory, lty = 2)
lines(h, predict(model))
legend("topleft", lty = c(1, 2), legend = c("Model", "Theory"))
```

With this model, we can now estimate the value for C_d by dividing the regression coefficient by the known constants.

```
model$coefficients[1] / (2 / 3 * sqrt(2 * g) * b)

I(h^(2/3))
 0.6123317
```

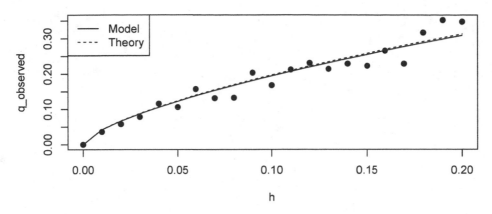

FIGURE 9.5 Regression of a polynomial function

9.7 Further Study

Although we are justified to conclude a positive relationship between contact frequency and hardship, this model is obviously too crude to draw conclusions about the customers' financial situation merely based on their contact frequency.

The powerful `lm()` function can manage much more than the simple models discussed in this chapter. Multiple linear regression is possible by adding a variable to the model specification with the plus symbol: `lm(hardship ~ contact + overdue, data = customers)`. This model assumes we have another explanatory variable called `overdue` that also explains hardship. Chapter 13 discusses multiple linear regression and in more detail.

Linear regressions are only one of the many methods to regress variables. The Generalised Linear Model function in R (`glm()`-) can model more advanced situations such as Poisson and binomial regressions. Poisson models measure variables that deal with counting things. You could, for example, model the number of phone calls made to a customer service centre. Binomial regression models deal with outcomes that are either true or false. Within the water context, an example of a binomial regression model could be predicting a pipe failure based on various condition variables such as pipe, material, asset age, and soil conditions.

10

Clustering Customers to Define Segments

The ideal form of customer service is personal attention, where the needs of each individual are met. Unfortunately, this level of service is more often than not impossible or too costly to achieve. Service providers, therefore, segment their customers into groups with similar characteristics. This approach allows service providers to better meet the needs of customers. Cluster analysis is a commonly used technique to segment customers using data. The primary objective of cluster analysis is to group observations based on their measured features. Cluster analysis divides an unlabeled data set into groups of observations with similar properties.

This chapter shows how to detect patterns and define segments in customer data. While factor analysis (Chapter 8) finds structure in the data to reduce the number of variables, cluster analysis detects structure in the data to classify groups of observations and thus adds an extra data point. This chapter shows two cluster analysis methods to segment customers based on simulated data. The learning objectives for this chapter are:

- Understand the principles of customer segmentation

- Apply and interpret hierarchical cluster analysis

- Apply and interpret k-means clustering

10.1 Customer Segmentation

Meeting the needs of every individual customer is the ultimate goal of providing commercial services. Unfortunately, almost no service provider can achieve this level of service, so they segment their existing and potential customer base into groups with similar needs and wants. Segmentation can include a wide range of approaches, such as (Kotler et al., 2010):

- *Demographic*: Age, gender, income, education, ethnicity (collected by national census)

- *Behavioural*: Purchasing habits, spending habits, user status, and brand interactions

- *Psychographic*: Interests, lifestyle, motivations, priorities

- *Geographic*: Town, postal code

This goal is almost impossible for water utilities as customers within a water system generally receive the same level of service. Water services to urban customers are standardised with limited options for individual differentiation. Customer segmentation is nevertheless a valuable exercise for water utilities. Defining segments allows a water utility to target communication about water savings and other useful bits of information to those customers who are most likely to benefit from it, rather than sending it to all customers. Segmentation also provides opportunities to offer special services, such as financial assistance programs to customers more likely to require this.

The information water utilities know about their customers is predominately behavioural and geographic. For example, water utilities know how much water each customer uses and thus how high their bills are. Utilities will also know whether a location is a household, shop, park, etc. Water utilities also know where customers live, which is particularly important when experiencing localised service issues, such as leaks or low pressure. With digital water meters (Chapter 11), water utilities can build even more detailed consumption patterns for fine-tuned segmentation.

Water utilities commonly don't know much about their customers' demographic and psychographic make-up. Demographic information is generally available from government census data, albeit highly aggregated. Psychographic information is generally derived from surveys about people's lifestyles and attitudes. It is the most insightful segmentation type but also the hardest to obtain. Specialised data brokers have developed trademarked statistical models that segment whole countries at the level of the household, such as Mosaic by Experian, VALS (Values and Lifestyles), and Values Segments by Roy Morgan. They obtain their data through various sources and use inferential analysis to assign a profile to individual households.

There are two approaches to segmenting customers. The top-down approach surveys a sample of customers and analyses this data to define the segments. The bottom-up approach uses existing data about individual customers to find clusters. The top-down approach is helpful for commercial organisations seeking to understand the market for potential and current customers. The bottom-up approach provides insights into the existing customer base to help an organisation serve them better. Both methods use the same statistical techniques.

Within the context of a water utility, all customers connected to the same network receive the same service. This fact does not mean that segmentation is not useful for utilities. Customer segmentation can help in preparing consumption profiles for more detailed network modelling and can help with better customer service when people contact the utility.

Water utilities have used the top-down method by sampling a group of water consumers (Burns et al., 2011; La & Vinot, 2008; Jenkins & Storey, 2012). The disadvantage of this approach is that it is difficult to ascertain which segment existing customers belong to. Water utilities are generally not researching new market opportunities, and to better service existing customers, your customer relationship management system needs to identify the most likely segment for each customer. The bottom-up approach is most valuable for water utilities because they are monopoly service providers and don't need to look for new customers. The mathematical and coding principles for both methods are the same. The main difference is that the data source is either a sample or a census. The additional complexity with only using a sample of customers is that the data needs to be extrapolated to the population of consumers to be useful.

This chapter discusses the two most commonly used clustering algorithms: hierarchical clustering and k-means. The examples in this chapter use small randomly generated data sets designed to illustrate the mathematical and coding principles behind cluster analysis. These principles can be easily applied to real-life data to create new insights into a group of customers.

10.2   Clustering Analysis Example

This simple example contains data from ten hypothetical customers (A–J). The first data dimension in the test data is the average annual water consumption, and the second is the size of the land on which the house resides.

The data consists of random numbers with a known distribution. The `rnorm()` function generates numbers with a normal distribution. Random numbers in a computer are not truly random, as they are calculated with an algorithm. The code is not further explained, and you can reverse-engineer it at leisure. This plot also introduces a new geometry. The `geom_label()` function shows a text at a coordinate with a little box around it (`geom_text()` plots without the box). Figure 10.1 intuitively shows that we should find two clusters.

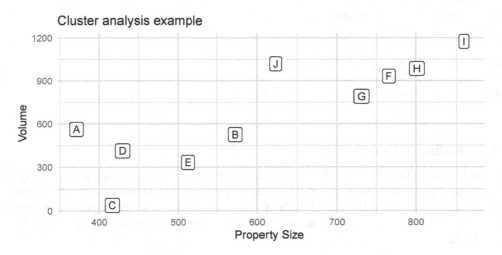

FIGURE 10.1 Clustering example.

10.3 Hierarchical Clustering

This method of clustering alliteratively defines clusters by identifying which data points are nearest to each other. In the first step, each data point is its own cluster. Each step identifies pairs of clusters by reviewing the distances between points and between clusters. The result is usually visualised in a dendrogram (tree diagram in Figure 10.4), which resembles an organisation chart, hence the hierarchical clustering.

Hierarchical clustering is the perfect analogy for customer segmentation. At the lowest level, each customer is their own segment. But for practical reasons, we must cut the tree at a manageable level of segments. Hierarchical clustering involves five steps:

1. Pre-process the data
2. Scale the data
3. Calculate the distances
4. Cluster the data
5. Review the outcome

10.3.1 Pre-Process the Data

For hierarchical clustering, the rows contain the items we want to cluster (customers), and the columns hold the variable we want to cluster by (customer characteristics). The data in the example is in the format we want it to be as the rows contain the clustering variable (customers A–F), and the columns (`property_size` and `volume`) are the features by which we seek to cluster the customers. The first five rows of the test data are:

```
head(consumption, n = 5)

# A tibble: 5 × 3
  id    property_size volume
  <chr>         <dbl>  <dbl>
1 A             371.   562.
2 B             573.   525.
3 C             417.    37.3
4 D             430.   413.
5 E             513.   332.
```

10.3.2 Scaling Variables

You need to normalise the data when the features are not on the same scale. In the example, the land size (m^2) and water consumption (m^3) have different units and should be scaled. The scale() function normalises data. The default setting of this function scales each element by subtracting the mean and dividing the result by the standard deviation of the sample. The long-form of scaling a variable is:

```
with(consumption, (volume - mean(volume)) / sd(volume))

[1] -0.3181039 -0.4204075 -1.7736949 -0.7326182 -0.9554980  0.7050841
[7]  0.3145340  0.8542333  1.3804265  0.9460446
```

In this example, the inputs for the scale function are the features of the customers data frame (excluding the first column, as this is an identifier). The output of the scale() function is a matrix with normalised observations. The output also shows the centres (the means), and the scales are the standard deviations. The center and scale function parameters control how the scaling is undertaken. The volume column matches the previous result.

```
consumption_scaled <- scale(consumption[, -1])
consumption_scaled

       property_size      volume
 [1,]    -1.35556945 -0.3181039
 [2,]    -0.20324283 -0.4204075
 [3,]    -1.09527888 -1.7736949
 [4,]    -1.02139061 -0.7326182
 [5,]    -0.54637047 -0.9554980
 [6,]     0.90443632  0.7050841
 [7,]     0.69648968  0.3145340
 [8,]     1.10315900  0.8542333
 [9,]     1.43262582  1.3804265
[10,]     0.08514144  0.9460446
attr(,"scaled:center")
property_size        volume
     608.2982      676.7456
attr(,"scaled:scale")
property_size        volume
     174.9241      360.5306
```

10.3.3 Calculating Distances

Hierarchical clustering requires a matrix with the distances between each of the customers. Several methods are available to calculate distances, two of which are discussed below and shown in Figure 10.2.

The most common method calculates the Euclidean distance using the famous Pythagoras formula. For n-dimensional data, the distance between p and q is:

$$D_{pq} = \sqrt{\sum_{i=1}^{n}(p_i - q_i)^2} \tag{10.1}$$

Another common method to determine the distance between two points is the so-called taxi-cab or Manhattan distance. This is the distance a taxi would take traversing through a city with a gridded street design. The taxicab distance is thus the sum of horizontal and vertical distances:

$$d = |p_1 - p_2| + |q_1 - q_2| \tag{10.2}$$

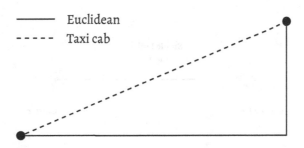

——— Euclidean

- - - - - Taxi cab

FIGURE 10.2 Euclidean and Taxicab distance.

The `dist()` function calculates the distance between the elements in a data frame or matrix. The function can use many methods, with Euclidean distance as the default. This function's output is a matrix the same size as the number of rows in the source data. In the example, we cluster ten customers, resulting in a ten-by-ten matrix with a multi-dimensional distance between them. However, the matrix only displays the lower triangle.

```
consumption_dist <- dist(consumption_scaled)
round(consumption_dist, 2)

       1    2    3    4    5    6    7    8    9
2   1.16
3   1.48 1.62
4   0.53 0.88 1.04
5   1.03 0.64 0.99 0.52
6   2.48 1.58 3.18 2.40 2.21
7   2.15 1.16 2.75 2.01 1.78 0.44
8   2.72 1.83 3.43 2.65 2.45 0.25 0.68
9   3.26 2.43 4.04 3.24 3.06 0.86 1.30 0.62
10  1.92 1.40 2.96 2.01 2.00 0.85 0.88 1.02 1.42
```

10.3.4 Clustering the Distance Matrix

Using the distance matrix, we can find the customer segments. The hierarchical clustering method iteratively creates groups of observations until all observations are in one cluster. In the example in Figure 10.1, customers F and G are the evident first cluster, as they have the smallest distance between them. The algorithm then looks for the next level, which consists of the cluster (F, G) and H. After four steps, clusters A–E and F–J fall into two clusters, and lastly, all customers are assigned to the supercluster.

The `hclust()` function takes a dissimilarity structure as the first argument. This is the same as a distance matrix produced by the `dist()` function. The default setting for hierarchical clustering is complete linkage, which means that the algorithm chooses the cluster or data point with the smallest maximum distance between points in sub-clusters. Several other methods are available with the `method` parameter. The most common ones are (Figure 10.3):

- Single linkage: minimum distance between two sets (`"single"`)

- Average linkage: distance between centres of gravity (`"average"`)

- Complete linkage: maximum distance between two sets (default)

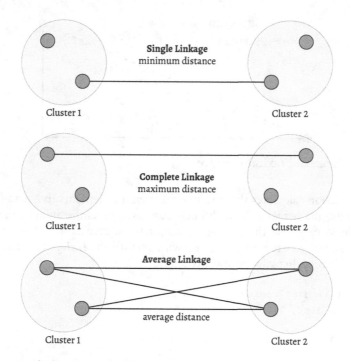

FIGURE 10.3 Common linkage types in cluster analysis.

```
customer_clusters <- hclust(consumption_dist)
```

The output of the `hclust()` function summarises the clustering results. When printing it in the console, R displays a summary. The output of the clustering function is a list with seven variables.

```
customer_clusters


Call:
hclust(d = consumption_dist)

Cluster method   : complete
Distance         : euclidean
Number of objects: 10
```

The best way to review the output is to plot it, which gives a dendrogram (tree diagram). The base plot function recognises the input as the result of hierarchical clustering and will visualise it as a tree. The main and sub options provide the title and subtitle to the plot. The labels option adds the names to the cluster numbers (Figure 10.4).

```
plot(customer_clusters,
    main = "Clustering Example",
    sub = "Simulated data", xlab = NA,
    labels = consumption$id)
```

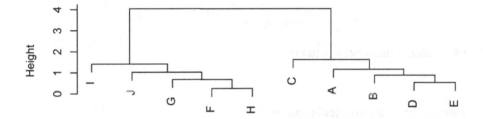

Clustering Example

Simulated data

FIGURE 10.4 Dendrogram of the example data.

You can view the clusters at each level of the analysis, working your way up to one supercluster. The vertical distance in the graph relates to the distance matrix and the linking method. The longer the line, the less related the segments are. Visually, both Figure 10.4 and 10.5 intuitively suggest that we should have two clusters.

You can extract more information from the clusters with the cutree() function. This function allows you to cut the tree at a certain level and effectively stop the clustering at that point. The output is a vector of the cluster number that each customer belongs to. At the highest level (k = 1), all customers form part of the same cluster. All customers are individuals at the lowest level (k = 10).

Extracting two clusters, we can assign these variables as segments to our customer table and visualise the data. Note the fill = factor(segment) option assigns a fill colour to the label. The factor() function forces R to assign qualitative colours instead of a numeric range. The scale_fill_manual() function defines the light and dark grey (Figure 10.5).

```
consumption$segment <- cutree(customer_clusters, k = 2)

ggplot(consumption, aes(property_size, volume, fill = factor(segment))) +
  geom_label(aes(label = id)) +
  scale_fill_manual(values = c("grey90", "grey60"), name = "Segment") +
  labs(title = "Customer Segmentation Example",
       subtitle = "Hierarchical Cluster Analysis",
       x = "Property Size", y = "Annual volume") +
  theme_bw(base_size = 10)
```

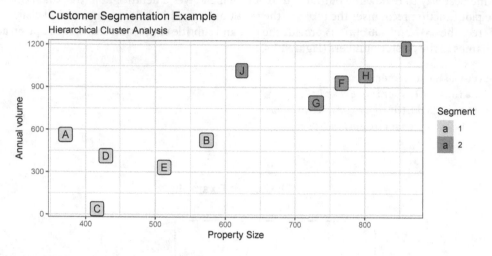

FIGURE 10.5 Clustered customer segments.

10.3.5   Interpreting Hierarchical Clustering

How do we know the ideal number of clusters? Two clusters solve this problem, but the boundary between clusters is not always this clear. Interpreting the results and selecting the ideal number of clusters combines practical insight and statistical analysis.

The diagram in Figure 10.5 helps to visually review the number of clusters. This simulated data is intuitively best fitted with two clusters. We can see this clearly because there are only two dimensions. When the number of dimensions exceeds four, visualising the data this way becomes almost impossible. The dendrogram in Figure 10.4 provides a better overview. The vertical line between one and two clusters in the dendrogram is the longest, which means the distance between the two clusters is larger than the distance between any other cluster.

Clusters can be intuitive but can also be easily misinterpreted. You can analyse any data and find clusters, but that does not imply that these clusters are meaningful and significantly distinct from other solutions. Some statistical techniques exist to assess how well the chosen model fits the data, but there is no single objective criterion to determine the ideal number of clusters.

Another method to visualise the clustering solution is the elbow method or scree plot, shown in Figure 10.6. This diagram visualises the distances between each of the cluster solutions so and identifies the largest jump in distance. The `height` variable in the clustering results list stores the height for each cluster. The order needs to be reversed with the `rev()` function to create a typical scree plot. You are looking for the point where the attached lines have the smallest internal angle, which in this case is number two, confirming the two-factor solution.

```
par(mar = c(4, 4, 1, 1))
plot(rev(customer_clusters$height), type = "b",
    xlab = "Clusters", ylab = "Distance")
abline(v = 2, lty = 2, col = "grey50")
```

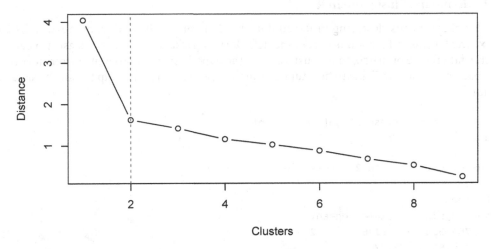

FIGURE 10.6 Scree plot of the customer clusters.

The elbow method or scree plot often doesn't provide a conclusive answer. The main criterion for the result of cluster analysis is that it makes sense. We strive for parsimony, meaning we want the lowest possible number of segments. The analyst's task is to work with the data user to find the optimal number. The main criterion is whether the cluster model explains the variability in customers observed in the field. A segment for each customer beats the purpose, and one customer segment is pointless. A manageable number that meets the purpose of the segmentation is ideal. This means you occasionally have to compromise between the statistical analysis and the practical purpose of the segmentation to find the perfect number.

10.4 *K*-means Clustering

The *k*-means algorithm is a stochastic method, which means that it uses an iterative approach to optimise the clustering solution. The algorithm uses the following steps:

1. Specify the number of clusters (*k*)
2. Randomly select objects from the data as the centroid of the initial *k* clusters
3. Assign each observation to its closest centroid using a distance matrix
4. Update the cluster centroids by calculating the mean values of all the data points in the clusters
5. Iterate steps 3 and 4 until the cluster assignments stop changing or the maximum number of iterations is reached

The main difference between these two methods is that with *k*-means the analyst has to specify the number of clusters to extract, which hierarchical clustering provides all possible solutions.

Another important difference is that the *k*-means method is stochastic, as it uses a random starting point. This means that the outcome is not always the same in complex situations. Lastly, the *k*-means method is much faster and can analyse large data sets, contrasting with hierarchical clustering.

10.4.1   *K*-means Clustering in R

To undertake *k*-means clustering in R, you don't have to normalise the data or create a distance matrix. The kmeans() function is part of the default stats package and takes at least two parameters: the data frame or matrix to be clustered and the hypothesised number of clusters (centers). Other parameters are available to fine-tune the analysis, explained in the help file for the kmeans() function.

```
k_clust <- kmeans(consumption[, -1], centers = 2)
k_clust

K-means clustering with 2 clusters of sizes 5, 5

Cluster means:
  property_size   volume segment
1      755.9989 979.6146       2
2      460.5974 373.8767       1

Clustering vector:
 [1] 2 2 2 2 2 1 1 1 1 1

Within cluster sum of squares by cluster:
[1] 108761.7 201013.8
 (between_SS / total_SS =  78.6 %)

Available components:

[1] "cluster"      "centers"      "totss"        "withinss"     "tot.withinss"
[6] "betweenss"    "size"         "iter"         "ifault"
```

The output provides a summary of the model being two clusters, which resulted in five observations each. The following section summarises the means of the clusters: The clustering vector shows which segment each observation has been assigned to. Next we see the Within-Cluster Sum of Squares for each cluster, which measures how close each observation is to the centre of its cluster, explained further in the next section. The output closes with a list of variables stored in the results.

10.4.2   Using the Elbow Method

As we have to choose the number of clusters in advance, we need a method to determine the optimal solution. We can do this by calculating the Within-Cluster Sum of Squares, which is the sum of the Euclidean distances between each observation and the centroid of that cluster.

The Within-Cluster Sum of Squares will decrease when the number of clusters increases. Calculating this metric for a series of clusters creates an elbow plot. Just as with hierarchical clustering in Section 10.3.5, the point where the decrease in value flattens (the smallest angle) is the most likely optimal solution.

The `kmeans()` function calculates the Within-Cluster Sum of Squares and provides it through the `tot.withinss` variable. The `for()` function loops through a vector to calculate the total for clusters one to nine. The result of the analysis is stored in the `within_ss` variable, used for plotting the elbow. The interpretation of these results follows the same logic as the hierarchical clustering example (Figure 10.7).

```
within_ss <- vector()

for(k in 1:(nrow(consumption) - 1)) {
  cl <- kmeans(consumption[, -1], centers = k)
  within_ss[k] <- cl$tot.withinss
}

par(mar = c(4, 4, 2, 1))
plot(within_ss, type = "b",
     xlab = "Clusters", ylab = "within-Cluster Sum of Squares")
abline(v = 2, lty = 2)
```

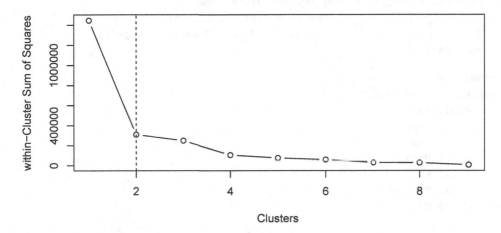

FIGURE 10.7 *K*-means Elbow Plot.

These two examples were simplistic as the solution is visually apparent. However, this is not the case when we work with more than three dimensions. We also need more advanced methods to deal with categorical data instead of numeric data. The next case study shows how to deal with such a situation.

10.5 Clustering Categorical Data

The example in this chapter uses continuous data for which it was easy to calculate the distance using the Euclidean or Taxicab distance. Customer surveys often contain categorical data, using boxes for the respondent to choose a predefined option. Calculating the distance matrix for this type of data requires some special considerations.

The customers that completed the survey in the second case study from previous chapters also completed some demographic questions about themselves. We also use the affective consumer involvement score calculated in Section 8.5. The data is stored in a separate CSV file in the data folder.

- household: Household type (Family, couple or single)

- garden: Does the property have a garden?

- ownership: Owning or renting?

- involvement: Affective involvement (score between 5 and 35)

```
segments <- read_csv("data/customer_segments.csv")
```

The affective involvement data is continuous instead of categorical. For consistency, we need to convert these numbers to a categorical variable. The cut() function can attach labels to a numeric variable by defining the breakpoints. The code below creates a new categorical variable for involvement with three labels instead of a number.

```
segments <- read_csv("data/customer_segments.csv")

segments$involvement_cat <- cut(segments$involvement,
                    breaks = c(5, 15, 25, 35),
                    labels = c("Low", "Medium", "High"))
head(segments[, 4:5])

# A tibble: 6 × 2
  involvement involvement_cat
        <dbl> <fct>
1          33 High
2           9 Low
3          25 Medium
4          28 High
5          22 Medium
6          35 High
```

Visualise all four segmentation variables in one *ggplot2* faceted bar chart to explore the content of this data.

10.5.1   Processing Categorical Variables

The problem with clustering categorical variables is when, for example, garden can take on the values "yes" or "no". If we simply encode these numerically as 1 and 2, then we effectively say having no garden is twice as much as having a garden. Direct encoding of categorical variables can lead to counter-intuitive outcomes. The standard method is to create dummy variables, which can take either zero or one.

In the case of a variable with two options, we can use a built-in function. The ifelse() function takes a test as the first argument, checking whether the variable garden contains "Yes". The next option in this function determines what happens if the result of the test is TRUE and, lastly, what happens when it is FALSE.

```
ifelse(segments$garden == "Yes", 1, 0)
```

The *fastDummies* package provides two functions to simplify this process using variables with more than two options (Kaplan, 2020). The `dummy_cols()` function splits any variable into its unique components, creating a new variable for each option. For the household variable, this function creates a new variable for each option and assigns a 0 or 1 to them. Of the columns that only have two options, one should be removed because otherwise, they correlate perfectly with each other. The numeric version of the involvement is also dropped.

```
library(fastDummies)
segments_dummy <- dummy_cols(segments,
                             remove_selected_columns = TRUE) %>%
  select(-involvement, -garden_No, -ownership_Renter)

glimpse(segments_dummy)

Rows: 400
Columns: 8
$ household_Couple      <int> 1, 1, 1, 1, 0, 0, 1, 0, 1, 0, 0, 1, 0, 1, 1, 0,...
$ household_Family      <int> 0, 0, 0, 0, 0, 1, 0, 1, 0, 1, 0, 0, 1, 0, 0, 0,...
$ household_Single      <int> 0, 0, 0, 0, 1, 0, 0, 0, 0, 0, 1, 0, 0, 0, 0, 1,...
$ garden_Yes           <int> 1, 0, 1, 1, 0, 1, 0, 1, 0, 1, 0, 0, 1, 1, 1, 1,...
$ ownership_Owner      <int> 1, 0, 0, 1, 1, 1, 1, 1, 1, 0, 0, 1, 1, 1, 0, 0,...
$ involvement_cat_Low  <int> 0, 1, 0, 0, 0, 0, 0, 0, 0, 0, 1, 0, 0, 0, 0, 0,...
$ involvement_cat_Medium <int> 0, 0, 1, 0, 1, 0, 1, 0, 1, 1, 0, 1, 0, 0, 1, 1,...
$ involvement_cat_High <int> 1, 0, 0, 1, 0, 1, 0, 1, 0, 0, 0, 0, 1, 1, 0, 0,...
```

With this transformed data, we can calculate the dissimilarity matrix, which is equivalent to a distance matrix. In the case of categorical variables, it makes little sense to speak about distances, so we need to ascertain the level of similarity between variables.

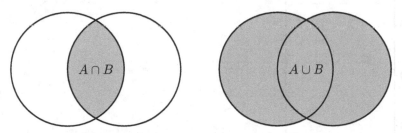

FIGURE 10.8 Jaccard similarity index.

The `dist()` function also implements the binary method, which uses the Jaccard index. This index expresses the ratio between the intersection of two variables ($A \cap B$) and the union of these variables ($A \cup B$). In other words, it is the ratio between the number of times both observations are true (1) and the number of variables, as shown in Figure 10.8.

$$J(A, B) = \frac{A \cap B}{A \cup B} \qquad (10.3)$$

The distance or dissimilarity between two observations is the reverse of the Jaccard index: $1 - J(A, B)$. The `dist()` function implements a version of this concept with the binary method. In this method, vectors are interpreted as binary bits, so non-zero elements are 'on' and zero elements are 'off'. The distance is the proportion of bits in which only one is on amongst those in which at least one is on.

The code below displays the distance matrix for the first ten rows of the data. Binary distance is the proportion of dummy variables in which *only one* is one (0 or 1), amongst those of which at least one is one (0 or 1 and 1 or 1). The code below displays the dissimilarity matrix for the first five rows and stores the complete matrix in a new variable.

```
round(dist(segments_dummy[1:5, ], method = "binary"), 2)
segments_dist <- dist(segments_dummy, method = "binary")

     1    2    3    4
2 0.80
3 0.60 0.75
4 0.00 0.80 0.60
5 0.83 1.00 0.80 0.83
```

10.5.2   Analysing Categorical Clusters

We can follow the same process as in the previous sections with this knowledge. The *factoextra* package contains several advanced methods and functions to assist cluster analysis (Kassambara & Mundt, 2020). This package includes a wealth of tools useful for multivariate analysis. For example, the `fviz_nbclust()` function visualises the elbow plot to assist with selecting the correct ideal number of clusters. Using the Within-Cluster Sum of Squares shows that three clusters provide a possible solution (Figure 10.9).

```
library(factoextra)
fviz_nbclust(segments_dummy, kmeans, method = "wss") +
  geom_vline(xintercept = 3, lty = 2)
```

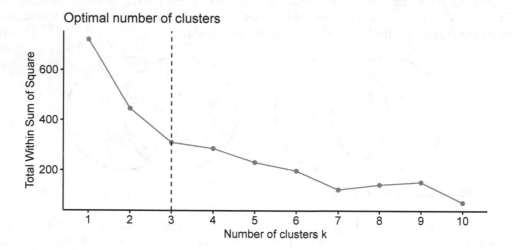

FIGURE 10.9 Elbow method with the factoextra package.

We can now evaluate the clusters with the *k*-means method using the distance matrix and extracting three clusters.

```
segments_k <- kmeans(segments_dist, centers = 3)
```

Visualising the results of multi-dimensional data is tricky. One way to do this is to combine multiple plots, showing all two-dimensional combinations of the variables and colouring the points by segment. This approach can be tedious as the number of variables increases. The image in Figure 10.10 shows one of the six possible combinations of segmentation variables.

```
segments$k <- segments_k$cluster

ggplot(segments, aes(household, involvement,
                     pch = factor(k))) +
  geom_jitter(size = 2, col = "darkgrey") +
  scale_shape(name = "Segment") +
  theme_minimal()
```

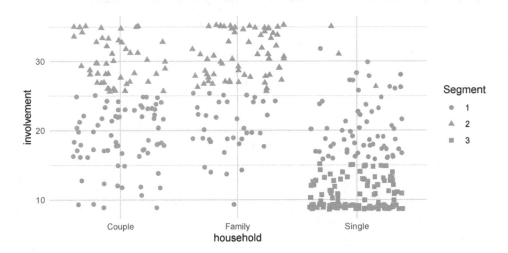

FIGURE 10.10 Cluster solution for household type and affective involvement.

The example in this section uses random data and a predetermined pattern, so the results are only examples of how to cluster categorical data. The code to create this synthetic data is available in Appendix C. The principles explained in this chapter can be easily extended to detect customer segments in more complex data.

10.6   Further Study

Cluster analysis is an ideal tool to classify customers into groups to enable a better level of service. Other applications for this analysis in the water utility context include spatial analysis, for example, finding clusters of complaints. John Snow's famous cholera map is the earliest example of this type of analysis. Snow mapped a cholera outbreak in London in the 1850s, which showed a cluster of disease around the pump in Broad Street. Clustering is also essential in sequencing the genes of pathogens in water. A dendrogram can also be used as an evolutionary tree of pathogen mutations. Dome cluster analysis techniques can be used to detect anomalies in multi-dimensional data, which is further discussed in Chapter 12.

This chapter only briefly discussed two of the most common clustering algorithms available in base R. Hierarchical and k-means clustering are the easiest methods to learn but are by no means the only ones. The four main types of commonly used methods are:

- Centroid: k-means clustering

- Connectivity: Hierarchical clustering

- Density: Clusters are areas of higher density than the remainder of the data

- Distribution: Objects belonging to the same statistical distribution

There is no objective criterion for selecting the correct algorithm. Each method has strengths and weaknesses, and choosing the best approach depends on the structure of the data and the purpose of the analysis. Several R packages are available to implement these methods. There is no single correct clustering method, as the interpretability of the outcome is more important than statistical sophistication. Generating clusters is a science, but interpreting them is an art.

11

Working with Dates and Times

Most data collected in operational processes are time series. These are data sets where each measurement is associated with the time of the measurement or observation. For example, SCADA systems, Internet-of-Things devices, and the laboratory data from case study 1 are time series. A time series is any type of data that contains repeated measurements of the same parameter over time. For example, this data could be continuous measurements of the pressure in the water mains, daily consumption, monthly performance or annualised financial results. Any data where the x-axis displays time and the y-axis a numeric value is a time series.

This chapter introduces the third case study, which looks at data collected from water meters from individual houses. When meter reads are collected at a high frequency, the data can provide valuable insights into water consumption, such as leak detection. This chapter explains how the R language manages dates and times and uses this knowledge to explore time series data. The learning objectives for this chapter are:

1. Understand the principles of how the R language processes date and time variables
2. Apply the functionality of the *Lubridate* package to manipulate dates and times
3. Use date and time variables in calculations and visualisations

11.1 Date Variables

Measuring time is as old as humanity. One of the first signs of civilisation is when a group of people creates a calendar to ensure they can organise themselves. Although we use time measurements effortlessly in our daily lives, the mathematical foundations are not that straightforward. For example, the earth's rotation around the sun does not neatly fit into our sense of numeric aesthetics of round numbers. Also, our time measurement method doesn't use decimal numbers but relies on an ancient sexagesimal (sixty-based) number system. Measuring time is a fascinating area where astronomy, anthropology, and mathematics meet, with some unique challenges when analysing data.

To reduce the complexity of measuring time, computers store times and dates as the number of days from a defined point of origin. R uses the Unix epoch, which starts counting on 1 January 1970. Other computer systems use different epochs. For example, the epoch for date and time variables in spreadsheets begins on 1 January 1900 or 1 January 1904, which is a frequent cause of confusion (Broman & Woo, 2018). The underlying data for all time and date variables in a Unix-based system like R is thus a large integer, counting days from the origin. Unix time also has a millennium problem. At 03:14:08 UTC on 19 January 2038, 32-bit versions of the Unix timestamp recycle to zero. Modern computers use 64-bit numbers, so there is enough space for a further 292 billion years before we experience another millennium problem.

The R function `Sys.Date()` displays the current date. The variable type is a date, which displays dates in a human-readable format, but underneath the display are the number of days in the Unix epoch, time zone, and other metadata. The `as.numeric()` function converts a variable to a number or results in NA when this is impossible.

```
Sys.Date()
as.numeric(Sys.Date())
```

```
[1] "2023-02-14"
[1] 19402
```

You can display dates in any arrangement with the `format()` function. The default date format is YYYY-MM-DD, which complies with ISO 8601, an international standard on dates and time representation (ISO 8601, 2019). R and other languages use a special syntax to change how it displays a date and time. For example `format(Sys.Date(), "%A %d %B %Y")` shows the current date as "Tuesday 15 June 2022". The codes start with a per cent sign, followed by an indicator. Some of the most commonly used ones are:

- %A: Name of the day

- %B: Full name of the month (%b for the abbreviated month)

- %d: Day of the month (1 – 31)

- %m: Number of the month (1 – 12)

- %Y: Year with century (%y results in the last two digits of the year)

To view a complete list of date conversion codes, read the `strptime()` help file. The format parameter can also include characters, such as commas, dashes, or whatever else you require. For example, `format(Sys.Date(), "%d/%m/%Y")` results in 19/06/2022.

11.1.1 Defining Date Variables

You can create a date variable by converting a character string or an integer with the `as.Date()` function. This function converts an ISO 8601 date character string (YYYY-MM-DD) into a date class variable. This function can also take dates in other formats, but you need to specify its structure with the `format` parameter. Lastly, this function can calculate a date using an integer and an origin, which can be either the Unix epoch or any other date. These examples all result in the same date variable.

```
as.Date("2022-07-01")
as.Date("1 July 2022", format = "%d %B %Y")
as.Date(19174, origin = "1970-01-01")
```

```
[1] "2022-07-01"
[1] "2022-07-01"
[1] "2022-07-01"
```

Date variables can also be used in calculations. The `as.Date()` and `Sys.Date()` functions provide the answer. The results of calculations with dates and times are of a special `difftime` variable class. To use this result in further calculations, you need to convert it to a numerical value using the `as.numeric()` function.

```
d <- Sys.Date() - as.Date("1969-09-12")
d
as.numeric(d)
```

```
Time difference of 19513 days
[1] 19513
```

11.2 Time Variables

The R language records the number of seconds since the Unix epoch to store a time variable. The current time is available with the `Sys.time()` function, which includes the date and time in the standard ISO 8601 formatting, for example, `2022-03-15 20:22:11 AEST`. The R language has no specific data type for storing a time, only a combination of date and time. You can change the format of a date/time variable in the same way as dates, using some additional modifiers. You can review additional options in the help entry for the `strptime()` function.

- `%H`: Number of hours (00–23)

- `%I`: Number of hours (01–12)

- `%M`: Minutes (00–59)

- `%p`: AM/PM indicator

- `%S`: Seconds (00–59)

- `%T`: Same as `%H:%M:%S`

```
format(Sys.time(), "%I %p")
```

```
[1] "07 AM"
```

11.2.1 Defining Time Variables

The `as.POSIXct()` function creates a date-time variable, using the same principles as with dates, for example:

```
as.POSIXct("2020-12-21 12:23:00")
as.POSIXct("21 December 2020 12:23", format = "%d %B %Y %H:%M")
```

```
[1] "2020-12-21 12:23:00 AEDT"
[1] "2020-12-21 12:23:00 AEDT"
```

The different time zones across the globe can be confusing, especially when confounded by daylight savings. For this reason, SCADA and IoT systems often provide time-series data in Co-ordinated Universal Time (confusingly abbreviated as UTC due to political compromise). This zone corresponds with the zero meridian over Greenwich without adjusting for daylight savings.

When you ask R for the current time, the time zone is displayed in a standard abbreviation. In my case, the time zone is AEST (Australian Eastern Standard Time). The `Sys.timezone()` function displays your computer's time zone. In addition, you can display any date-time variable in another time zone with the `format()` function.

```
Sys.timezone()
Sys.time()
format(Sys.time(), tz = "UTC")
format(Sys.time(), tz = "NZ")
```

```
[1] "Australia/Melbourne"
[1] "2023-02-14 07:09:33 AEDT"
[1] "2023-02-13 20:09:33"
[1] "2023-02-14 09:09:33"
```

The `format()` function *displays* a time zone as a character variable but does not change the underlying time variable. Unix time is always counted from UTC, and the underlying integer never changes to maintain consistency and prevent analytical jet lag. To set the time zone, you need to change the time zone attribute of the variable. The code below converts the current system time to New Zealand time.

R uses the standard IANA (Internet Assigned Numbers Authority) time zones. These use a consistent naming scheme, typically in the form of `<continent>/<city>`. There are a handful of exceptions because not every country lies on a continent. Some designations are abbreviations, such as "CEST" for Central-European Summer Time. To find other time zone designations, review the output of the `OlsonNames()` function, named after Arthur David Olson, who compiled the first version of this list.

```
ams <- Sys.time()
ams

attr(ams, "tzone") <- "Europe/Amsterdam"
ams

[1] "2023-02-14 07:09:33 AEDT"
[1] "2023-02-13 21:09:33 CET"
```

To create a variable with a specific time zone, you need to add the `tz` option to the declaration.

```
as.POSIXct("21 December 2020 12:23",
           format = "%d %B %Y %H:%M",
           tz = "America/Vancouver")

[1] "2020-12-21 12:23:00 PST"
```

When converting a POSIX time variable to date, you can run into a quirk of the R language. Even though the time is displayed in the local time zone, R interprets it as being in UTC when converting it to a date variable. To prevent this issue, you can add `tz = ""` to enforce the local time zone. R assumes that the POSIXct time value supplied to `as.Date()` is in the UTC time zone, which is how R stores all POSIXct values. For example, the code below shows an Australian date-time variable displayed as yesterday's date in UTC.

```
d <- as.POSIXct("2022-02-22 10:00:00")
d
as.Date(d)

[1] "2022-02-22 10:00:00 AEDT"
[1] "2022-02-21"
```

The R language has extensive capabilities to manage time zones and convert between them. However, when working with time-critical data, you should be vigilant that your analysis is presented in the correct time zone. The default time zone is UTC, which makes it easier to work between time zones but can also cause confusion because your computer operates in local time.

11.3   The Lubridate Package

The *Lubridate* package forms part of the Tidyverse and provides syntactic sugar (simplified functions) to work with date and time variables (Grolemund & Wickham, 2011). This package uses the same underlying R data types but provides some additional functionality and lubricates some tasks of the basic R date and time functions.

```
library(lubridate)
```

The *lubridate* package has some useful functions to extract components of a data and time variable. For example, the year(), month(), and day() functions extract the number of the year, month, or day. The now() function is a *lubridate* alternative for Sys.time().

```
now()
year(now())
month(now())
day(now())
```

```
[1] "2023-02-14 07:09:33 AEDT"
[1] 2023
[1] 2
[1] 14
```

For grouping measurements by a period, it is often useful to round a date-time variable to the nearest day, week, month, or other periodicities. The floor_date(), round_date() and ceiling_date() functions provide flexible options to round a date-time variable. Date-time variables can be rounded by second, minute, hour, day, week, month, bi-month, quarter, season, "halfyear" or year by using these strings as options.

```
floor_date(now(), unit = "hour")
round_date(now(), unit = "day")
ceiling_date(now(), unit = "month")
```

```
[1] "2023-02-14 07:00:00 AEDT"
[1] "2023-02-14 AEDT"
[1] "2023-03-01 AEDT"
```

Lubridate simplifies working with time zones with the with_tz() and force_tz() functions. The with_tz() function converts a date-time variable to the new time zone, while the force_tz() function only changes the time zone name without changing the underlying variable.

```
now()
with_tz(now(), tz = "Europe/Amsterdam") # Change display
force_tz(now(), tz = "Europe/Amsterdam") # Change variable
```

```
[1] "2023-02-14 07:09:33 AEDT"
[1] "2023-02-13 21:09:33 CET"
[1] "2023-02-14 07:09:33 CET"
```

The code in the remainder of this book extensively uses *lubridate* functionality without explaining it in detail. The functions are self-explanatory, and I leave it to the reader to reverse-engineer the code. You can read the *lubridate* vignette to learn more about the functionality of this package: vignette("lubridate").

11.4   Exploring Digital Metering Data

Traditionally water utilities measure consumption from individual customers when generating a bill. As manual reading water meters is a time-consuming task, this activity occurs at most once every month. This limited information is suitable for sending bills, but it does not reveal anything about the consumer's patterns of consumption. This approach is like looking at a monthly bank balance and then figuring out where the money went.

Current technology enables water meters to read and transmit data at a much higher frequency so that we can gain deeper insights into how consumers use their water. This data unlocks a raft of benefits, from providing customers with detailed knowledge of their consumption to leak detection and network optimisation.

This case study uses the functionality of the Tidyverse and some specialised packages to analyse smart meter data. The term smart meters is common but contains a bit of marketing spin. Most customer water meters in this category are standard devices fitted with an electronic data logger and transmitter. These water meters are not intrinsically smart, but they provide the utility with detailed data that allows water professionals to make smarter decisions.

Digital water meters measure and transmit data at varying frequencies, from every second to daily reads. Deciding how much data to collect depends on the use case. Water engineers typically prefer a reading every five minutes to match their modelling data frequency, while the billing department would be more than happy with one daily reading. The customer service team likes to know whether a property has leaks, which requires at least hourly reads. Each use case has its own ideal data collection frequency. While a high data collection frequency would satisfy everybody's needs, there is also a high cost. The higher the data rate, the transmission bandwidth, reduced battery life and data storage costs. In addition, collecting data at high frequencies is potentially unethical because it reveals too much information about customers' lifestyles (see Section 1.4.4). Hourly reading seems a good compromise because it enables most of the sought benefits, limits privacy risk for customers, and is within reasonable reach for the current level of technology.

Digital metering data is a time series because each data point represents a measurement at a given location at a point in time. The data in this case study is an equally-spaced time series, which means that all time intervals are the same, besides some exceptions when data was not transmitted. The data for this case study is synthetic and based on assumptions and stochastic variables because using actual data could lead to a privacy breach. The `meter_reads.csv` file in the `data` folder contains the simulated digital metering data with three fields:

- `device_id`: Unique id for the data logger

- `date_time`: Date and time of the reading

- `count`: Number of revolutions. Each revolution equals five litres

```
library(readr)
meter_reads <- read_csv("data/meter_reads.csv")
```

The data contains one year of meter reads in the distant future of an imaginary city from one hundred simulated customers. Some meter reads are unavailable due to telemetry issues. We can use this data to analyse the effectiveness of the telemetry network and water consumption.

11.4.1   Filtering and Grouping by Date and Time

Although each device transmits every hour, the telemetry network is never perfect and can miss certain transmissions, depending on local circumstances. For example, some meters might be behind a wall or in a pit, or atmospheric conditions prevent the signal from reaching the base station.

The `transmissions` tibble contains a count of the daily transmissions for all devices. Note that these dates are in the UTC time zone.

```
library(dplyr)
transmissions <- meter_reads %>%
  mutate(date = as.Date(timestamp)) %>%
  count(device_id, date)
transmissions
```

```
# A tibble: 10,000 × 3
   device_id date            n
       <dbl> <date>      <int>
 1   1006044 2050-01-01      9
 2   1006044 2050-01-02     11
 3   1006044 2050-01-03      9
 4   1006044 2050-01-04     13
 5   1006044 2050-01-05     13
 6   1006044 2050-01-06     11
 7   1006044 2050-01-07      8
 8   1006044 2050-01-08      8
 9   1006044 2050-01-09     12
10   1006044 2050-01-10     10
# ... with 9,990 more rows
# i Use `print(n = ...)` to see more rows
```

Filtering a tibble by date is as easy as filtering other values. Although the `as.Date()` conversion is not always necessary, it is best to always include it to avoid confusion. The `between(x, left, right)` function is a shortcut for `x >= left & x <= right`. Note that a date variable is not the same as a date-time variable. When filtering a date-time variable by a specific date, you need to convert it to a date variable.

```
filter(transmissions, date == as.Date("2050-02-04"))

filter(transmissions, between(date, as.Date("2050-01-01"), as.Date("2050-01-31")))

filter(meter_reads, as.Date(timestamp) == as.Date("2050-02-04"))
```

You can also use helper functions from the *lubridate* package to zoom in on the data. The first line of code below filters all data for the 2050 calendar year. The second line filters for all entries in March 2050, and the last line filters the raw data for all transmissions within a certain hour as the `round_date()` function rounds all time stamps to their nearest hour. The last line shows how to achieve similar results without using the *lubridate* package.

```
filter(transmissions, year(date) == 2050)

filter(transmissions, floor_date(date, "month") == as.Date("2050-03-01"))

filter(meter_reads, round_date(timestamp, "hour") ==
                    as.POSIXct("2050-03-01 10:00:00", tz = "UTC"))

filter(meter_reads, format(timestamp, "%b %Y") == "Mar 2050")
```

The floor, round, and ceiling functions within the *lubridate* package also provide a flexible toolkit to group a data frame by the defined time windows. The table below shows the average number of data transmissions received each calendar month. The `days-in-month()` is a helper function from the *lubridate* package to calculate the mean number of transmissions.

```
transmissions %>%
  group_by(month = floor_date(date, "month")) %>%
  summarise(transmissions = sum(n)) %>%
  mutate(days = days_in_month(month),
         mean_transmissions = transmissions / days)

# A tibble: 4 × 4
  month       transmissions  days mean_transmissions
  <date>              <int> <int>              <dbl>
1 2050-01-01          71097    31              2293.
2 2050-02-01          64268    28              2295.
3 2050-03-01          71092    31              2293.
4 2050-04-01          22952    30               765.
```

11.4.2 Analysing Water Consumption

To determine the level of consumption between two reads in litres per hour, we need to subtract two consecutive reads from each other. A fast way to determine the difference between successive numbers in a vector is the `diff()` function, which we saw in Section 4.4.1. However, if we apply this function to the whole data frame, we get in trouble moving from one device to the next. Taking the last read of one meter and the first of the following meter results in negative consumption.

The *dplyr* package provides the `lag()` and `lead()` functions to compute differences over the rows in a vector. These functions shift the vector by one item by default. For example, `lag(1:3)` results in `NA`, `1`, `2` and `lead(1:3)` results in `2`, `3`, `NA`. The first and last results are unavailable because there is no value before the first one or after the last one. The `n` parameter in this function sets the number of positions to lead or lag by. This function can be easily used with grouped data frame.

The example below calculates the volume for each meter read as the differences between two consecutive measurements by subtracting the lagging value and multiplying the difference by five to convert a count to litres. The second part calculates the time difference. Given the missing values, we cannot assume it is always one hour. This method results in `NA` values for the flow for the first hour for each water meter, so they are filtered out. Lastly, the `ungroup()` function undoes the grouping, so we have a clean data set suitable for further analysis.

```
meter_flow <- meter_reads %>%
  group_by(device_id) %>%
  mutate(volume = (count - lag(count, default = 0)) * 5,
         time = as.numeric(timestamp - lag(timestamp)),
         flow = volume / time)  %>%
  filter(!is.na(flow)) %>%
  ungroup()
```

11.4.3 Linear Interpolation to Calculate Daily Flows

Calculating daily volumes requires that we know the flow at the end of each day. However, these digital meters don't necessarily provide a reading at midnight. So, we need to interpolate the data to ensure we have a reading at midnight for each water meter.

The approx() function interpolates between data points. By default, this function undertakes a linear interpolation for fifty equally spaced points. This function can also undertake constant interpolation, which selects the value of the previous point until a new point is available. The n parameter specifies the number of interpolated points. You can also specify for which x-values you like to find the interpolated value. The result of this function is a list with the x and y vectors.

The interpolation function converts date-time variables to their numeric equivalent, which needs to be corrected when displaying the results. The code below visualises these options of this method (Figure 11.1).

```
x <- today() + c(0, 7, 10, 15)
y <- c(0, 30, 135, 200)

linear <- approx(x, y)
constant <- approx(x, y, method = "constant", n = 9)
point <- approx(x, y, xout = today() + 8.5)

plot(as.Date(linear$x, origin = "1970-01-01"),
     linear$y, col = "gray", cex = .5,
     xlab = "x", ylab = "y")

points(constant$x, constant$y, col = "gray", pch = 20, type = "b")
points(x, y, col = "red", pch = 16)
points(point$x, point$y, col = "blue", pch = 12, cex = 2)
```

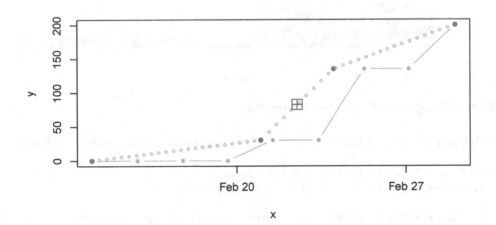

FIGURE 11.1 Basic workings of the approx() function.

Let's apply this function to the digital metering data. The first step creates the times for which we like to determine the flow, which in our case is midnight of every day. The code then groups the meter reads by device and applies the linear interpolation. Finally, the last step calculates the daily volume and cleans the data by removing NA values and undoing the grouping. The histogram in Figure 11.2 shows that some water meters show a very high consumption, which is caused by leaks (see further Chapter 12).

```
daily_dates <- unique(floor_date(meter_reads$timestamp, "day"))

daily_flow <- meter_reads %>%
  group_by(device_id) %>%
  summarise(date = (daily_dates),
            count = approx(x = timestamp,
                           y = count,
                           xout = daily_dates)$y) %>%
  mutate(volume = 5 * (count - lag(count))) %>%
  filter(!is.na(volume)) %>%
  ungroup()

library(ggplot2)
ggplot(daily_flow, aes(volume)) +
  geom_histogram(binwidth = 100) +
  theme_minimal()
```

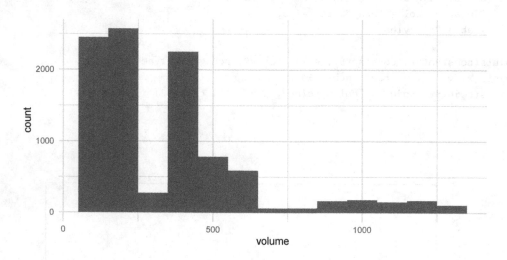

FIGURE 11.2 Histogram of daily water consumption.

Another method to create date sequences is with the `seq.Date()` and `seq.POSIXct()` functions.

```
seq.Date(from = Sys.Date(), length.out = 10, by = 1)
seq.POSIXt(from = Sys.time(), length.out = 4, by = 30)
```

```
[1] "2023-02-14" "2023-02-15" "2023-02-16" "2023-02-17" "2023-02-18"
[6] "2023-02-19" "2023-02-20" "2023-02-21" "2023-02-22" "2023-02-23"
[1] "2023-02-14 07:09:34 AEDT" "2023-02-14 07:10:04 AEDT"
[3] "2023-02-14 07:10:34 AEDT" "2023-02-14 07:11:04 AEDT"
```

11.4.4   Diurnal Curves

The most informative method to visualise water consumption is through a diurnal curve. A diurnal cycle occurs every rotation of the earth. For water consumption by households, water consumption during the night is zero or close to zero, with a morning peak and a second peak around dinner time. The shape of the curve for each household depends on the number of people in the dwelling and their lifestyle. The diurnal curve tells a story about the people who live in this house. The diurnal curve of an individual connection should be treated with care as this data reveals a lot about people's lifestyles and is thus privacy-sensitive. Diurnal curves also apply to whole systems when we aggregate the data over a group of water meters, often called a District Metered Area (Cominola et al., 2019).

To calculate the diurnal curve over all connections, we need the flow rate of each customer at whole hours. However, each device transmits at a random offset, which is unlikely to be a whole hour. We could undertake tens of thousands of interpolations, which is computationally intensive, especially in real-life situations with millions or even billions of data points. The least computationally intensive method is to round each reading to its nearest hour. For each individual meter, it will cause an error, but aggregated over all meters, the positive and negative errors (meters transmitting before and past the whole hour) cancel each other out.

The expression() function creates mathematical expressions that allow for using subscripts, superscripts, and other mathematical constructions. The plotmath() help file contains details on the conventions used to plot mathematical notation.

```
diurnal <- meter_flow %>%
  mutate(timestamp_au = with_tz(timestamp, tzone = "Australia/Melbourne"),
         hour = hour(round_date(timestamp_au, "hour"))) %>%
  group_by(hour) %>%
  summarise(min_flow = min(flow) / 1000,
            mean_flow = mean(flow) / 1000,
            max_flow = max(flow) / 1000)

library(ggplot2)
ggplot(diurnal, aes(x = hour, ymin = min_flow, ymax = max_flow)) +
  geom_ribbon(fill = "gray", alpha = 0.5) +
  geom_line(aes(x = hour, y = mean_flow), col = "black", size = 1) +
  ggtitle("Connections Diurnal flow") +
  ylab(expression(Flow~m^3/h)) +
  theme_bw(base_size = 10)
```

The graph in Figure 11.3 tells a story about the people who live in the homes represented in this data. They have their morning showers around 7 am, most are home during the day and start to

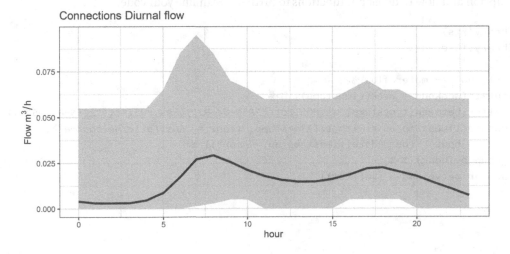

FIGURE 11.3 Cumulative diurnal curve over all properties.

go to bed after 8 pm. When looking at an individual property, you can work out how often they flush their toilets in the early hours of the day. This shows that this type of data is privacy-sensitive when looking at individual properties. Aggregation over multiple properties is one way to reduce this risk. Also in this graph we see the influence of the leaking properties as the maximum flow is much higher than the average.

You can easily adjust this code snippet to filter between dates or for one or more connections. Filtering for a series of connections is made easier with the `%in%` function. The two function calls below provide the same results.

```
filter(meter_flow, device_id %in% c(1404857, 1515776))

filter(meter_flow, device_id == 1404857 | device_id == 1515776)
```

11.5   Further Study

Working with time series is an essential skill for any data scientist. Besides using a data frame with a timestamp variable, R also has a specialised time series variables. This data class provides specialised functions such as decomposing a time series into a seasonal and a random component.

The code example below creates a simple forecast, extrapolating the last few days of the available data. The data frame is first converted to a time series object with the XTS package, which provides extended time series functionality (Ryan & Ulrich, 2022). The Forecast package contains the `auto.arima()` function which applies an Auto-Regressive Integrated Moving Average (ARIMA) model to the time series to enable forecasting future values (Hyndman & Khandakar, 2008). The plot in Figure 11.4 shows the available data and the predicted values, with the level of uncertainty indicated by the shaded areas.

These models analyse the patterns of the data in the past to predict the future. ARIMA modelling is unaware of the causes of the level of consumption as it repeats patterns of the past, which is only useful for short-term predictions. The model in this code is only reliable within twelve hours after the last data point. ARIMA is unable to take changing circumstances into account, such as changing attitudes to water consumption.

The next chapter delves deeper into the digital metering data and looks for anomalies in consumption and how to develop R functions to further streamline your code.

```
library(xts)
library(forecast)

hourly_vol <- meter_flow %>%
  mutate(weeknum = week(timestamp)) %>%
  filter(between(timestamp, as.POSIXct("2050-04-01"), as.POSIXct("2050-04-05"))) %>%
  mutate(timestamp_au = with_tz(timestamp, tzone = "Australia/Melbourne"),
         hour = round_date(timestamp_au, "hour")) %>%
  group_by(hour) %>%
  summarise(volume = sum(flow))
```

```
hourly_vol_ts <- xts(hourly_vol$volume,
                     order.by = hourly_vol$hour,
                     format = "%Y%m%d")

fit <- auto.arima(hourly_vol_ts)
fcast <- forecast(fit, h = 12)
plot(fcast)
```

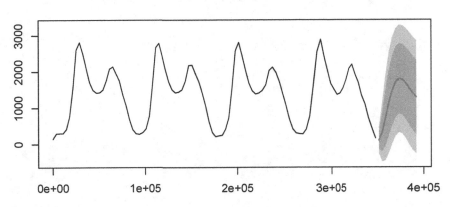

Forecasts from ARIMA(2,0,1) with non−zero mean

FIGURE 11.4 Short-term forecast of future water consumption.

12

Detecting Outliers and Anomalies

Organisations often have more data than they can manually process. Reporting all this data has little value because it overwhelms the organisation with information that is not actionable. Actionable reports tell operational staff or management that something is different from usual. Developing data dashboards filled with traffic lights and dials might be satisfying aesthetically and impressive to the uninitiated, but it can lead to false confidence. A more productive approach to processing the data deluge is to show those processes where something interesting has occurred. A range of techniques is available to detect anomalies in the data. Reporting outliers and anomalies focusses the organisation's attention by raising questions and motivating action. This chapter discusses finding the most exciting points in your data to create actionable reports. This chapter also shows how to further streamline your code with functions and develop a leak detection tool to use with digital metering data. The learning objectives for this session are:

- Apply statistical methods to detect outliers

- Find anomalies in a time series

- Develop R functions to streamline your code

- Write a function to detect leaks from digital metering data

12.1 Detecting Anomalies

The massive deluge of data is a boon for organisations, but it has the disadvantage that we often don't know where to look for actionable insights. Detecting anomalies helps focus data consumption on those aspects of a service process that require attention. Unfortunately, the terms outlier and anomaly are often used interchangeably, and there is no agreed definition in the data science literature. Other terms used in the literature are abnormalities, discordant, or deviants (Aggarwal, 2016).

Detecting deviant points is essential in assessing the quality of the data. When detecting an outlier, either the measurement instrument was not calibrated, or there were issues during storage and transmission. Anomalies also occur in cases where data is entered manually into a database. Detecting these anomalies is essential to prevent wrong conclusions.

Even more interesting are outliers caused by the underlying process. Outlier detection helps to detect suspicious monetary transactions, diseases in medical imaging, or even signals from extraterrestrial intelligence. Outliers are also important in water management. For example, outliers in the turbidity of a filtration process can increase the risk of pathogens in the water and threaten public health. Also, detecting anomalies in billing data can identify customers who might be struggling to pay their bills. A good grasp of anomalies helps managers and operational staff to focus.

Generally, an anomaly or outlier is a single observation or a collection of data points that do not follow the same pattern as the rest of the data. This description reveals a selective judgement.

We first need to know what normality looks like to detect an abnormality in data. The difference between standard data, noise, and anomalies is not always clear. While noise in the data consists of meaningless variations caused by natural variance, outliers are points of interest. Determining the boundary between these two categories is often a matter of judgement that requires domain expertise.

The most straightforward outlier-detection method uses predefined anomalies. For engineering processes, government regulations and technical standards often define the limits of a process. We used this technique in Chapter 4 when comparing laboratory results with regulatory limits. Anomalies can also be defined by the type of measurement. For example, any detection of zero pH can be safely removed from the data. Domain knowledge is, therefore, an essential skill in assessing the validity of anomalies.

However, we don't always know where the boundary between outlier and normal data lies. Statistical techniques help to define this boundary.

12.1.1 Graphical Detection

Visualising data is a logical starting point to detect anomalies in observations. Outliers are often easily seen when looking at a time series or a histogram. The human mind has evolved to detect patterns in our environment and determine which observations are different than the others. Most visualisations, such as a time-series graph or histogram, show outliers. Boxplots are the ideal visualisation geometry for visually detecting abnormalities in data (Section 4.7.2). The R algorithm for boxplots marks values smaller or larger than the first recorded value below 1.5 times the IQR from the 25th and above the 75th percentile with a circle. The first part of this chapter uses the laboratory data from the first case study (Figure 12.1).

```
library(readr)
library(dplyr)
library(ggplot2)

chlorine <- read_csv("data/water_quality.csv") %>%
  filter(Measure == "Chlorine Total")

ggplot(chlorine, aes(Suburb, Result)) +
  geom_boxplot()
```

The problem with visualising anomalies is that a human operator needs to review the data manually and find the irregularities. Reviewing graphs is a time-consuming activity and is also prone to biases. A problem with human anomaly detection is that the mind is so good at detecting patterns that we see them even though there is nothing to see. So why not let a robot do the work for you?

Statisticians and data scientists have developed an extensive suite of tools to detect anomalies, some of which are described below. The most apparent outliers in any data are the minimum and maximum values. But the highest and lowest values are not necessarily anomalies. Statistical techniques can detect measurement outliers by comparing them to a theoretical distribution.

12.1.2 Standard Deviations

A standard method to identify anomalies is to find every observation more than three standard deviations away from the average. The outlier detection method in linear regression models discussed in Section 9.5.2 uses this method. Applying this concept to the chlorine data for Merton, we find two anomalies, the highest of which is the maximum value.

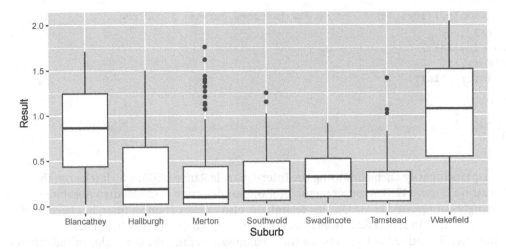

FIGURE 12.1 Example of a boxplot with outlier detection.

In the example below, this function identifies the values greater than three standard deviations from the mean. The `which()` function gives the index number of the values that meet the condition of being an outlier. There are two further functions to find extreme values in a vector. The `which.min()` and `which.max()` provide the index number of either the lowest or highest value of a vector. We can use the results of the which-type functions to subset a data frame.

```
cl_merton <- filter(chlorine, Suburb == "Merton")

outliers <- which(abs(cl_merton$Result - mean(cl_merton$Result)) /
                  sd(cl_merton$Result) >= 3)

cl_merton[outliers, c("Sample_No", "Sample_Point", "Result", "Units")]

# A tibble: 2 × 4
  Sample_No Sample_Point Result Units
      <dbl> <chr>         <dbl> <chr>
1    617952 ME_19428       1.76 mg/L
2    660877 ME_14915       1.62 mg/L
```

The threshold of three standard deviations is a subjective assessment. A value of three means that 0.13% of the data is an outlier given a normal distribution. The analyst can choose different threshold values, should the context require this.

Using standard deviations from the mean is only effective when the data is close to normally distributed, which is rarely the case for water quality data. Moreover, outliers impact both the mean and the standard deviation. This indicator is thus problematic, as it defines outliers and, at the same time, the outliers influence the indicator itself (Leys et al., 2013).

12.1.3 Median Absolute Deviation

Using the median to detect outliers offers the advantage over the mean of being insensitive to the presence of outliers (Leys et al., 2013). The new mean will approach infinity when adding an infinite value to a vector, but the median changes little, as shown below.

```
x <- sample(1:100, 10) # Ten random integers between 1 and 100
mean(x)
mean(c(x, Inf))
median(x)
median(c(x, Inf))
```

```
[1] 68.7
[1] Inf
[1] 75
[1] 76
```

Boxplots identify anomalies using the Inter-Quartile Range (IQR), which relies on the median observation (Figure 12.1). Another method for outlier detection is the Median Absolute Deviation (amusingly abbreviated as MAD). This technique, also called a Hampel filter, has been applied to detect anomalies in sensor data from stream flow measurements (Bae & Ji, 2019).

The MAD is defined as 1.4826 times the median value of the absolute value of the differences between each observation (x_i) and their median ($\tilde{x}$). The multiplication factor applies to normal distributions. For other distributions, use one divided by the third quartile, which for a normal distribution is 1.4826 (Leys et al., 2013).

$$\text{MAD} = 1.4826 \times \text{med}(|x_i - \tilde{x}|) \tag{12.1}$$

```
1.4826 * median(abs(cl_merton$Result - median(cl_merton$Result)))
mad(cl_merton$Result)
```

```
[1] 0.111195
[1] 0.111195
```

The distribution of water quality data is rarely normally distributed, so we need to correct the MAD by using the reciprocal value of the third quartile as the constant.

```
cl_p75 <- quantile(cl_merton$Result, 0.75, names= FALSE)
cl_mad <- mad(cl_merton$Result, constant = 1 / cl_p75)
cl_mad
```

```
[1] 0.1724138
```

The MAD can act as an alternative to the standard deviation and detect anomalies using the same approach described in the previous section. In addition, the MAD method leads to a lower threshold for outlier status than standard deviations, as the data has a strong positive skew.

```
median(cl_merton$Result) + 3 * cl_mad
```

```
[1] 0.6172414
```

By creating a new variable to identify the outliers, we can visualise them (indicated by a circle in Figure 12.2). This code introduces the formula method to plot data, previously used in linear regression in Chapter 9.

```
outliers <- which(abs(cl_merton$Result - median(cl_merton$Result)) / cl_mad >= 3)
cl_merton$outlier <- FALSE
cl_merton$outlier[outliers] <- TRUE

par(mar = c(4, 4, 1, 1))
plot(Result ~ Date, data = cl_merton, pch = (!outlier) + 20, type = "b")
```

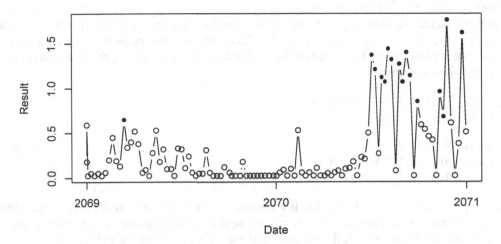

FIGURE 12.2 Anomalies detected with the Median Absolute Deviation.

12.1.4 Grubb's Test

The techniques discussed previously are deterministic in that the analyst has to decide the multiplication factor that defines deviations. More advanced methods undertake hypothesis testing to detect anomalies.

Grubb's Test evaluates whether the data point furthest from the mean is an outlier by comparing it to a t-distribution. This test only works for a single point within normally-distributed data. The null hypothesis for this test is that there are no outliers, so any p-value lower than the critical value (usually 0.05) means that the maximum or minimum value is an outlier.

The *outliers* package contains a collection of functions to test data for outliers, including Grubb's Test (Komsta, 2022). This function includes some parameters to control the calculations. Adding `opposite = TRUE` tests whether the lowest value in the data is an outlier.

```
outliers::grubbs.test(cl_merton$Result)
```

```
Grubbs test for one outlier

data:  cl_merton$Result
G = 3.47105, U = 0.88527, p-value = 0.01919
alternative hypothesis: highest value 1.76 is an outlier
```

In the output, G indicates how far the highest value deviates from the mean. The additional value U is the ratio of sample variances, with and without the suspicious outlier. The output of the function in this example indicates that the samples' highest measured level of chlorine is an outlier as the p-value is below 0.05.

12.1.5 Time Series Anomalies

The methods for finding statistical outliers described in the previous section detect one or more observations that deviate from the general pattern of the rest of the data. Still, these points will not necessarily be related over time. An example of a collective anomaly is a heat wave, which the World Meteorological Organisation defines as five or more consecutive days of prolonged heat in which the daily maximum temperature is higher than the average maximum temperature by 5 °C or more.

The `rle()` function, an acronym for Run-Length Encoding, summarises a vector into its observation frequencies. The run-length of a vector is a count of how often a value repeats. So the run lengths for `c(7, 23, 23, 43, 43, 43)` is 1, 2, 3. The `rle()` function helps to compress a data set by removing duplicates. This function can also find spikes in data. Let's apply this function to create a plain-vanilla spike detector.

```
runlength <- rle(cl_merton$Result > 1)
runlength

Run Length Encoding
  lengths: int [1:11] 80 2 1 4 1 4 9 1 3 1 ...
  values : logi [1:11] FALSE TRUE FALSE TRUE FALSE TRUE ...
```

The input of the run-length function is a Boolean vector of chlorine results over 1 mg/l. Using this information, you can visualise those parts of the time series that show an anomaly, ensuring that operators can focus on critical data instead of manually reviewing multiple reports.

If we assume that an outlier is defined by four or more consecutive detections of greater than 1 mg/l, then we need to do some additional work. The `rep()` function repeats the first parameter by its second parameter. In the example below, all run-length sections where the value is TRUE (result greater than 1 mg/l) and the length is larger than four are marked as anomalies, which we can then visualise in Figure 12.3.

```
cl_merton$outlier <- rep(runlength$lengths >= 4 & runlength$value,
                         runlength$lengths)

ggplot(cl_merton, aes(Date, Result, col = outlier)) +
  geom_line(col = "gray") +
  scale_x_date(date_labels = "%b %Y") +
  geom_point(size = 2) +
  scale_color_grey(start = 0.8, end = 0.2, name = "Outlier") +
  theme_bw() +
  theme(legend.position = "bottom")
```

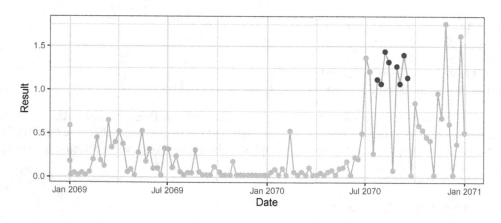

FIGURE 12.3 Run-length outlier detection.

12.1.6 Managing Outliers

The first task after detecting anomalies is to report these to a human operator for review. When using data with anomalies for further analysis, you will have to decide whether to remove, impute, or leave the outliers. The same caution should be used when removing or imputing outliers, as with missing data (Section 7.7). Outliers should only be removed when they are an issue of data quality, not when they communicate something about the process.

12.2 Extending R with Functions

The art of writing computer code is to create elegant and efficient code that is easy to understand and works as efficiently as possible. As a general rule, if you need to write the same code more than twice, there must be a more efficient method. This is because it takes time to write repetitive code; more importantly, duplicated code increases the risk of bugs.

Writing code in the console is fleeting and not recommendable for complex investigations. The next step of automation is to write a script and store it for future use. Chapter 7 discussed reusing pipes and scripts to avoid repetition. But there are further steps we can take to streamline code. R is a functional programming language, so you can easily extend its functionality. The next level of R coding is thus to write bespoke functions.

A function is a script that performs a specific task and can be used multiple times. You can do anything with functions that you can do with vectors:

- Assign them to variables

- Store them in lists

- Pass them as arguments to other functions

- Create them inside functions

To solve the questions in Chapter 2, you repeated the Kindsvater formula (Equation 2.1) multiple times. Functions help to alleviate this problem, and they have the added benefit that you can share them with other people to make their life easier. This example provides a custom function to calculate channel flow that uses $C_d = 0.62$ and $g = 9.81$ as sensible defaults:

```
channel_flow <- function(h, b, cd = 0.62, g = 9.81) {
    q <- (2/3) * cd * sqrt(2 * g) * b * h^(3/2)
    return(q)
}

channel_flow(h = c(100, 75, 120) / 1000, b = 0.5)

[1] 0.02894809 0.01880234 0.03805325
```

You assign the function to a variable with the function() syntax. The parentheses identify the parameters that the function uses. The channel_flow() function uses three variables: h, b, cd and g. A function does not need to have parameters. The default value of the discharge constant d is set at 0.62 and g at 9.81 m/s^2. The code of the function is placed between curly braces. This function defines the variable q and returns that value as the result of the function. The codes between curly braces can be as complex as you like and can call other functions.

When you evaluate this function, R expects you to either explicitly assign the variables or keep them in the same order as specified. However, the variables whose values are defined in the function definition do not have to be specified in the function call. You can also override the default parameters. For example, if, hypothetically speaking, you would be building an open water channel on Mars, the gravitational acceleration is $3.72 \, \text{m/s}^2$.

```
channel_flow(h = 100 / 1000, b = 0.5, cd = 0.62, g = 9.81)
channel_flow(100 / 1000, 0.5)
channel_flow(100 / 1000, 0.5, g = 3.72)
```

```
[1] 0.02894809
[1] 0.02894809
[1] 0.01782612
```

12.2.1 Functional Programming

R is a functional programming language which means that it provides tools for the creation and manipulation of functions. The majority of functions in the R language are themselves R scripts. You can view the content of any function by evaluating it without parentheses. The displayed byte-code is the memory address location for the function.

```
channel_flow
```

```
function(h, b, cd = 0.62, g = 9.81) {
    q <- (2/3) * cd * sqrt(2 * g) * b * h^(3/2)
    return(q)
}
<bytecode: 0x55cffa190cf0>
```

Not all functions are written in the R language. A small set of primitive operations, such as sum(), is compiled into machine language for greater computation speed. When you type sum without parenthesis, and press enter, R shows the function summary:

```
sum
```

```
function (..., na.rm = FALSE)  .Primitive("sum")
```

All other functions are written in the R language itself. This means that you can inspect the source code of almost all functions in the R language, which is useful when you need to find out exactly how the analysis is undertaken. If, for example, you like to know how R creates a boxplot, then you can review the code of the boxplot.stats() function. Reviewing this function could be the starting point for developing your own bespoke version of the boxplot. This feature demonstrates the power of open-source software, as it promotes the freedom of the user to inspect and modify the code.

12.2.2 Variables in Functions

R stores the content of the variables you declare by default in the global environment, which is stored in memory. An R session can have more than one environment. Each package has its own environment, and variables within a function are in their own environment, so they don't modify anything existing in the main environment. The environment tab in RStudio shows the global environment by default. This approach is called lexical scoping, illustrated by the code below.

The first code snippet defines a value for cd but does not pass it to the function. The function will thus use the version available in its environment instead of the one in the global environment.

```
cd <- 12
channel_flow(100 / 1000, 0.5)
```

```
[1] 0.02894809
```

By default, functions cannot modify variables in the global environment. If you seek to change a global variable from within a function, then you need to use the «– operator, as shown below.

```
q <- 0
channel_flow(100 / 1000, 0.5, g = 3.72)
q
```

```
[1] 0.01782612
[1] 0
```

```
q <- 0
channel_flow2 <- function(h, b, cd = 0.62, g = 9.81) {
  q <<- (2/3) * cd * sqrt(2 * g) * b * h^(3/2)
  return(q)
}
channel_flow2(100 / 1000, 0.5, g = 3.72)
q
```

```
[1] 0.01782612
[1] 0.01782612
```

12.3 Detecting Anomalous Water Consumption

Let's get back to the water consumption case study. Detecting anomalies in the flow of water helps to better manage this precious resource. Most water consumers display a predictable pattern of water consumption. Any abnormality in these patterns will, as such, reveal either a change in consumer behaviour or issues with their plumbing. Leaks are an example of an anomaly that is easily detected in digital metering data. However, abnormalities can also occur when customers leave a garden hose on, use evaporative air conditioners, or for any other justified reason.

12.3.1 Leak Detection

The basic principle of leak detection is that a water meter provides us information about what happens to the system downstream of the water meter. Indoor water consumption in residential properties is highly predictable, as it reflects the daily rhythm of sleeping, waking, working, and resting. The model diurnal curve for residential properties (Figure 12.4) has one or more periods without water consumption (Gurung et al., 2014).

People sleep at least once, or houses are unoccupied during the day. Customers will only notice substantial leaks when they are presented with their bills. Leak detection as a service by water utilities saves water but also prevents bill shock for customers (Monks et al., 2021). While detecting leaks might present a short-term reduction in water sales for the water utility, there are long-term benefits of reduced future investments in new assets.

This assumption implies that for any property where the minimum flow rate was not zero, the inhabitants were either using water continuously over that period, or there was another reason for

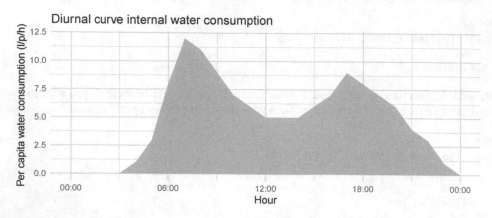

FIGURE 12.4 Model diurnal curve.

this flow unrelated to human behaviour, such as a leak. The longer the period the minimum flow is calculated for, the more confident you can be that the anomaly is a leak. This decision rule only applies to residential properties as commercial locations might have a justified reason for using water continuously over a long period.

To find a property with a leak, we must work out the lowest flow between two consecutive points over the data of interest and select those water meters with a minimum flow of more than zero.

```
leaks <- meter_reads %>%
  group_by(device_id) %>%
  mutate(time = as.numeric(timestamp - lag(timestamp)),
         volume = 5 * (count - lag(count, default = 0)),
         flow = volume / time) %>%
  summarise(min_flow = min(flow, na.rm = TRUE)) %>%
  filter(min_flow > 0)
```

This method is suitable for wrapping it into a function so we can do regular checks on leaking properties without repeating the code. The function in the next code snippet takes a vector of meter reads for a single meter and a vector of timestamps as input and presents the lowest minimum flow. If that flow is greater than zero, then we have a reason to believe there was a leak in their plumbing.

```
detect_leaks <- function(reads) {
  tibble(reads) %>%
    mutate(volume = 5 * (reads - lag(reads, default = 0))) %>%
    summarise(min_volume = min(volume, na.rm = TRUE)) %>%
    pull(min_volume)
}

detect_leaks(meter_reads$count[meter_reads$device_id == 1515776])
detect_leaks(meter_reads$count[meter_reads$device_id == 1873453])

[1] 30
[1] 0
```

Whether to take action on this information depends on business considerations. Not every leak is worth fixing, as the plumbing costs could far exceed the ongoing price of the water.

12.4   **Further Study**

This chapter described how to undertake univariate and time series anomaly detection using statistical techniques. Several methods for multivariate anomaly detection are available, such as *k*-Nearest Neighbours (KNN) and the Local Outlier Factor (LOF).

The nearest neighbour algorithm determines how close a point is to its *k* nearest neighbours using a method similar to cluster analysis, discussed in Chapter 10. Whether a point is an outlier depends on the distance from its nearest neighbours. The left panel of Figure 12.5 visualises the KNN approach. The size of the dots corresponds with their distance to the five (*k*) nearest points. The grey points are anomalies, as their distance is larger than a threshold. The *FNN* package provides fast nearest neighbours algorithms and applications (Beygelzimer et al., 2022).

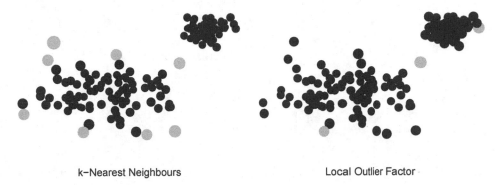

k−Nearest Neighbours   Local Outlier Factor

FIGURE 12.5 Simulation of *k*-Nearest Neighbours and Local Outlier Factor algorithms.

Other methods use point densities to classify a point as an outlier. For example, the LOF algorithm calculates the average density around the *k*-nearest neighbours divided by the average density around the point. The right panel of Figure 12.5 visualises the LOF approach, which shows that the LOF method recognises local outliers missed by the kNN algorithm. The *dbscan* package provides a library of density-based algorithms, including LOF (Hahsler et al., 2019).

Writing functions is a valuable tool to streamline your code and efficiently process data. You can share functions with colleagues to ensure analysis consistency and save them some time. The next level of development in the R language would be to write a package. An R package consists of a library of documented functions. The functions in a package need to be documented and provide a meaningful error message when a user provides invalid input. You can write packages to make your coding experience easier or develop a corporate package that codifies house styles, database connectors, and other standard functions. Several methods are available to share a package, such as adding it to CRAN.

13

Introduction to Machine Learning

Humans learn from the past by recognising patterns. But our ability to detect patterns is limited to small amounts of data and by natural biases. Machine learning is an approach to statistical analysis whereby a computer detects and then uses the results to predict outcomes from new data. The combination of large amounts of available data becoming cheaply available and open-source machine learning algorithms is causing a revolution in many industries, including water management. For example, water utilities can apply machine learning to predict which water or sewer main is likely to fail soon, determine future water demand to plan their assets, or assess which customers are most likely to struggle to pay their bills. This chapter introduces the principles of machine learning and the basic modelling process.

The learning objectives for this chapter are:

1. Understand the principles of machine learning
2. Apply cross-validation to linear regression
3. Implement a decision tree prediction

13.1 What Is Machine Learning?

Machine learning is a group of algorithms that convert the input into output by recognising patterns in data. Arthur Samuel from IBM coined this term in 1959 to promote their capabilities in software development (Burkov, 2019). Machine learning is a branch of Artificial Intelligence (AI), but it is not the only method to mimic human reasoning.

The family tree of machine learning includes a wide range of algorithms, with three main branches: supervised and unsupervised methods and reinforcement learning (Figure 13.1). Reinforcement learning is an advanced neural network technique made famous by AlphaGo, the first program to beat a world champion at the game of Go.

13.1.1 Unsupervised Machine Learning

Unsupervised machine learning methods train a computer using information that is neither classified nor labelled by humans. These algorithms trawl through the data to find patterns and help to discover otherwise invisible patterns in the data.

Clustering methods detect related groups in multi-dimensional data. A clustering algorithm predicts which category a point belongs to by analysing its relationship to other data points. Chapter 10 used this technique to segment customers.

Single vector decomposition, principal component analysis, and factor analysis (Chapter 8) are common unsupervised techniques to reduce the number of dimensions in data. These algorithms analyse the relationships between variables to test whether they can be combined into fewer variables. This reduction is beneficial because optimising the number of features (predictor variables) reduces the complexity of the problem.

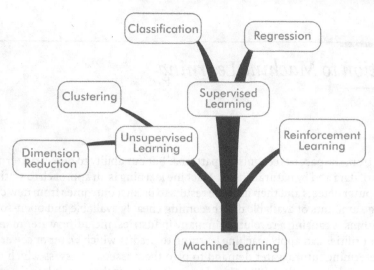

FIGURE 13.1 Overview of machine learning methods (Prevos, 2019).

The outlier detection techniques discussed in Chapter 12 are simple techniques to find anomalies using statistics. Some more advanced methods, such as *k*-Nearest Neighbour (kNN) Local Outlier Factor (LOF), and Isolation Forest, can detect abnormalities in multi-dimensional data. These algorithms use the similarity between points, which is also the primary input for clustering methods.

13.1.2 Supervised Learning

In supervised methods, the algorithm training data includes known relationships. When, for example, we predict how likely a sewer main will collapse, the training data consists of verified information about which pipes have collapsed and the circumstances under which this occurred, such as pipe material and age and soil, and surface conditions. A supervised algorithm analyses the training data to discover patterns to predict how likely a pipe will fail using new data. Supervised learning can classify data into groups or regress data to find relationships between variables and predict observations or measurements.

Recognising cracks in sewer pipe inspection videos is an example of a supervised classification problem. The algorithm analyses thousands of labelled images, some of which contain cracks and some do not. Human operators add these labels manually, so the algorithm learns how to recognise a crack using these examples. The algorithm can then detect failures in new footage, saving countless hours of watching footage of pipe inspections. These applications use neural networks for image recognition.

Supervised learning is similar to how humans learn to classify their lived experience. For example, a lecturer teaches engineering students how to classify damaged sewer mains, after which they independently review the inspection footage, even though new cases might arise.

Regression analysis finds relationships between numerical or categorical variables using known dependent and independent variables. Chapter 9 discussed simple linear regression, which is one application of this technique. Regression analysis is the most common form of machine learning. It is a form of supervised learning because the algorithm results in an equation that describes the relationship between the dependent and independent variables using known connections. Later in this chapter, we discuss a more complex regression analysis example.

13.1.3 Basic Principle of Machine Learning

The first step in any machine learning project is to ask the right question and find appropriate data. The next step involves selecting the predictor variables, also called features, that the algorithm will use. The algorithm then evaluates these features in a subset of the available data, called the training set. The outcome of this step is a model to classify or predict new observations. Finally, to determine how well the model performs, the prediction model is used on another part of the available data for evaluation, known as the testing or validation set.

13.2 Concrete Strength Case Study

Concrete is the most common material in the built environment. It is ideally suited due to its durability and high compressive strength. Concrete is also essential in water management as it is commonly used for pipes, tanks, and dam walls.

The strength of concrete is defined by the amount of pressure it takes to collapse a standardised cube, measured in MPa (Megapascal). Engineers predict the strength of concrete based on its components, which in its simplest form are cement, sand, gravel, and water. After the concrete is mixed, its strength gradually increases. Each mixture results in different concrete strength. Engineers also add other components, the so-called admixtures, such as blast furnace slag, fly ash, and plasticisers to improve the concrete (Chaubey, 2020).

The UCI Machine Learning Repository contains extensive data sets used to develop predictive models. It also includes data with over a thousand concrete mixtures and their compressive strength.[1] This data was initially used in a paper on using neural networks to predict compressive strength for a given mixture by (Yeh, 1998). This file contains the following variables:

- Cement (kg): Binding agent

- Coarse Aggregate (kg): Gravel

- Fine Aggregate (kg): Sand

- Water (l)

- Blast Furnace Slag (kg)

- Fly Ash (kg)

- Plasticiser (kg)

- Curing Age (days)

- Concrete compressive strength (MPa)

You can download the spreadsheet directly from the UCI website. The download.file() functions does exactly what it says. The destfile parameter defines the location and name of the downloaded file.

```
url <- "http://archive.ics.uci.edu/ml/machine-learning-databases/concrete/
        compressive/Concrete_Data.xls"
download.file(url, destfile = "data/concrete-data.xls")
```

---

[1]http://archive.ics.uci.edu/ml/datasets/Concrete+Compressive+Strength

The variable names in the raw data contain a lot of characters that complicate using the data in code, so they need some cleaning. The first step is to convert the variable names to lower case letters with the `tolower()` function. The `str_remove_all()` function shortens the variable names by removing punctuation with the `[[:punct::]]` regex code. Next, spaces are replaced by underscores. This leaves long variable names, so the next line removes `_component` and everything after it (`.*`). The `str_trim()` function removes spaces at the start or end of a string. These lines of code are only examples of how to clean text strings. You could, of course, manually set the names for this data frame.

The functions starting with `str_` originate from the stringer package that provides functionality to transform character stings (Wickham, 2022b). You can learn more about this package with `vignette("stringr")`. The `regular-expressions` vignette explains the principles of using regex to filter text variables.

```
library(readxl)
library(tidyverse)

concrete <- read_excel("data/concrete-data.xls")
names(concrete) <- tolower(names(concrete))
names(concrete) <- str_remove_all(names(concrete), "[[:punct:]]")
names(concrete) <- str_remove(names(concrete), "component.*")
names(concrete) <- str_trim(names(concrete))
names(concrete) <- str_replace_all(names(concrete), " ", "_")
names(concrete)[9] <- "compressive_strength"
glimpse(concrete)

Columns: 9
$ cement               <dbl> 540.0, 540.0, 332.5, 332.5, 198.6, 266.0, 380.0, ...
$ blast_furnace_slag   <dbl> 0.0, 0.0, 142.5, 142.5, 132.4, 114.0, 95.0, 95.0,...
$ fly_ash              <dbl> 0, 0, 0, 0, 0, 0, 0, 0, 0, 0, 0, 0, 0, 0, 0, 0, 0...
$ water                <dbl> 162, 162, 228, 228, 192, 228, 228, 228, 228, 228,...
$ superplasticizer     <dbl> 2.5, 2.5, 0.0, 0.0, 0.0, 0.0, 0.0, 0.0, 0.0, 0.0,...
$ coarse_aggregate     <dbl> 1040.0, 1055.0, 932.0, 932.0, 978.4, 932.0, 932.0...
$ fine_aggregate       <dbl> 676.0, 676.0, 594.0, 594.0, 825.5, 670.0, 594.0, ...
$ age_day              <dbl> 28, 28, 270, 365, 360, 90, 365, 28, 28, 28, 90, 2...
$ compressive_strength <dbl> 79.986111, 61.887366, 40.269535, 41.052780, 44.29...
```

This case study aims to develop models to predict the compressive strength of the concrete, using the amount for each component as a feature. As each mixture's compressive strength is known, we need to implement supervised learning techniques.

13.3   Cross-Validation

Just like human learning uses exams to test to what extent new knowledge has been retained, machine learning uses cross-validation to test the accuracy of models. Cross-validation splits the available data into training, testing, and sometimes validation subsets. Data scientists use the training data to develop the model and apply the mode to the testing and validation sets to confirm whether the model works on new data. Cross-validation tests the ability of the model to predict outcomes with data that was not used in developing the model.

Failing to use cross-validation will most likely result in a model with a high level of bias or variance. In machine learning, a model with low bias accurately predicts new observations. Bias is a concept similar to validity discussed in Section 8.1.1. Another error that can arise in machine learning is high variance. A high level of variance means that the model is sensitive to small fluctuations in the data (Burkov, 2019), which is analogue to a model's reliability.

When a model is not optimised, it is either underfitting or overfitting the data (Figure 13.2). Underfitting means that the model cannot accurately predict outcomes with new variables. So it has a high bias but low variance. Underfitting can occur when the model is too simple or needs more predictive variables. Overfitting occurs when the model perfectly follows the training data but cannot predict any new observations. A model that overfits the data has a low bias but a high variance. Overfitting can occur when using too many predictive variables, also called features.

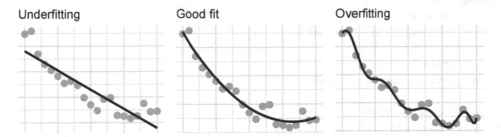

FIGURE 13.2 Fitting regression models.

A range of methods is available to implement cross-validation, the most common of which are:

1. Sub-sampling (hold out method): Create training and a testing set
2. k folds: Sample k sets for training and testing (usually 5–10)
3. Leave p out: Leave p observations from the data as the test set

Training data is usually around 80% of the available data, with the remainder used for testing. The last two methods are used iteratively to develop models for different pairs of training and testing sets. The parameters for the model are then combined to provide an optimal estimation, or the best-performing model is selected. For testing a model with k folds, the process is:

1. Split the training data into K equal parts
2. Fit the model on $k - 1$ parts and calculate test error using the fitted model on the kth part
3. Repeat k times, using each data subset as the test set once

The leave p out method is an extreme case of k folds, where k equals $n - (p - 1)$, where n is the number of rows in the data and p the number of observations used for testing. The sub-sampling method is another special case of k folds, where $k = 1$. For a time series, the folds should contain consecutive rows, and the training data needs to be directly after the testing data so that the order of observations remains intact.

It is important to remember not to use the testing data to develop the model and only use each testing set for one model version. If the same testing data is used to build or test multiple models, it effectively becomes training data, and the model can show high variance. The strict separation between training and testing data can be problematic when not much data is available, so use your data wisely.

13.4   Multiple Linear Regression

The linear regression model in Chapter 9 used a simple model to find the relationship between financial hardship and contact frequency. The concrete mixture data set is an ideal case study demonstrating how to implement multiple linear regression with cross-validation. This example uses basic sub-sampling to create training and testing sets. The code below takes a sample of 80% of the available data by selecting random row numbers with the `sample()` function. As we are building two models for this data, we need to split the testing set in two because we can only use each testing set once to validate a prediction.

```
set.seed(123)
n <- nrow(concrete)
train_id <- sample(1:n, 0.7 * n)

concrete_train <- concrete[train_id, ]
concrete_test <- concrete[-train_id, ]

t <- nrow(concrete_test)
test_id <- sample(1:t, 0.5 * t)

concrete_test1 <- concrete_test[test_id, ]
concrete_test2 <- concrete_test[-test_id, ]
```

We now build the models using the training data and validate them with the testing sets. Don't be tempted to explore the data in the testing sets, as it can introduce bias. To explore the data, we plot the relationship between compressive strength and each of the components of the concrete mixture (Figure 13.3) using the training data. The first line in the mutate function replaces all underscores with spaces, and the second line converts the text to title case to clean the labels in the graph.

```
pivot_longer(concrete_train, -compressive_strength,
             names_to = "component", values_to = "amount") %>%
  mutate(component = str_replace_all(component, "_", " "),
         component = str_to_title(component)) %>%
  ggplot(aes(amount, compressive_strength)) +
  geom_point(alpha = .3, col = "darkgray") +
  facet_wrap(~component, scales = "free_x") +
  geom_smooth(method = "lm", se = FALSE, col = "black", linetype = 5) +
  theme_minimal() +
  labs(x = "Amount (kg)", y = "Strength (MPa)")
```

Figure 13.3 shows that no single parameter can accurately predict compressive strength. Due to the competing relationships, mixing concrete to achieve the required compressive strength is a complex problem. A multiple linear regression could provide a better model. This section compares two models to predict compressive strength using the mixture variables.

The example below models the relationship between the water/cement ratio, coarse and fine aggregate, and curing time. The -1 parameter in the formula sets the intercept as zero because a mixture with any of these essential components is not concrete, so that is a justified choice. The `I()` (identity) function is a method to combine predictor variables. You need to use this function because most arithmetic symbols mean something else in formula notation. The identity function

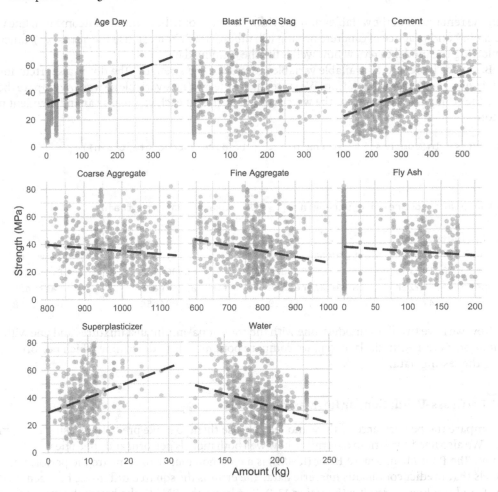

FIGURE 13.3 Relationship between components and compressive strength for the training data.

brings them back to their original meaning. Read the help file of the `formula()` function for more details.

The output of this multiple linear regression shows the model's coefficients in the same output as with simple linear regressions we saw in Chapter 9. Note that wrapping a line with an assignment operator in parentheses will also print the result without needing an additional line of code.

```
(lm_min <- lm(compressive_strength ~ I(water / cement) +
                coarse_aggregate + fine_aggregate + age_day - 1
            , data = concrete_train))


Call:
lm(formula = compressive_strength ~ I(water/cement) + coarse_aggregate +
    fine_aggregate + age_day - 1, data = concrete_train)

Coefficients:
 I(water/cement)   coarse_aggregate    fine_aggregate        age_day
      -22.43231            0.03096           0.02282        0.10127
```

The output shows that the water-cement factor is the strongest predictor for concrete strength, with the remainder of the variables only providing a small contribution. This outcome is in line with the theory and practice of concrete mixing (Chaubey, 2020). You can use the `summary()` function or plotting the model to review how well it fits the training data.

But what if we add all available variables to predict the strength? The dot is a shortcut for all variables in the R formula notation. We store this model in a new variable so we can compare them. Note that this model doesn't use the water/cement ratio but each variable as an independent predictor.

```
(lm_all <- lm(compressive_strength ~ . - 1, data = concrete_train))


Call:
lm(formula = compressive_strength ~ . - 1, data = concrete_train)

Coefficients:
          cement  blast_furnace_slag               fly_ash              water
        0.107158            0.090874              0.066361          -0.151749
superplasticizer    coarse_aggregate        fine_aggregate            age_day
        0.473149            0.011730              0.004456           0.121698
```

Now we have two linear models, one with the traditional minimalist mixtures and one with the various admixtures. In the last step, we compare both models to determine which performs best using the testing data.

13.4.1 Cross-Validation for Regression Models

To compare the performance of these models, we use them to make predictions using the testing data. We also need a metric to determine how well both models perform and continue with the best option. The Root Mean Square Error (RMSE) is a common metric to compare the performance of models that predict continuous numeric data. The error is the squared difference between prediction $\hat{y}$ and the known result y (Equation 13.1). The lower the RMSE, the better the model fits the data, with zero indicating a perfect match between prediction and reality.

$$RMSE = \sqrt{\frac{(\hat{y} - y)^2}{n}}$$

(13.1)

The modelling functions in R all result in a similar output that includes the equations that define the relationship between observed variables. The `predict()` function uses the result of a model produced with R and runs that model over the new data set. The result of this function is a vector of predicted values, in this case, concrete strength. If the function has no `newdata` parameter, it uses the training data.

```
concrete_test1$predict <- predict(lm_min, newdata = concrete_test1)
concrete_test2$predict <- predict(lm_all, newdata = concrete_test2)
```

A quick function avoids repetition when calculating the RMSE for multiple models.

```
rmse <- function(yhat, y) {
  sqrt(mean((yhat - y)^2))
}


(rmse_min <- rmse(concrete_test1$predict, concrete_test1$compressive_strength))
(rmse_all <- rmse(concrete_test2$predict, concrete_test2$compressive_strength))
```

```
[1] 14.46851
[1] 10.86919
```

Using the RMSE, the model with all parameters outperforms the model with only the basic ones. We can also verify this difference visually by graphing how well the prediction compares with the measured data in the testing set (Figure 13.4). The diagonal in the two graphs indicates a perfect match between reality and prediction. The closer the points align to the diagonal, the better the model can predict the outcomes. If all points are perfectly aligned with the diagonal, then $RMSE = 0$.

```
par(mfcol = c(1, 2))
plot(predict ~ compressive_strength, data = concrete_test1,
    main = "Parsimonious model", pch = 20,
    sub = paste("RMSE =", round(rmse_min)))
abline(a = 0, b = 1, col = "red")
plot(predict ~ compressive_strength, data = concrete_test2,
    main = "All variables", pch = 20,
    sub = paste("RMSE =", round(rmse_all)))
abline(a = 0, b = 1, col = "red")
```

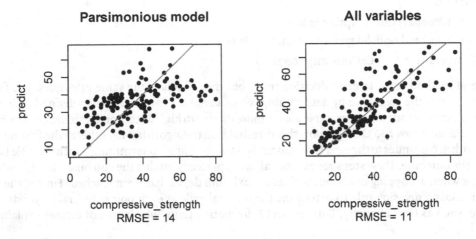

FIGURE 13.4 Visual comparison of concrete strength models.

13.5 Decision Trees

The last example in this chapter introduces decision trees, another form of supervised machine learning. A decision tree is a classification algorithm that categorises data using a series of features, the predictive variables. We have already seen cluster analysis (Chapter 10), a collection of unsupervised classification algorithms because the method finds related observations using mathematical rules based on distance or similarity. In a decision tree, a human has classified a data set, and the algorithm then finds the relationship between the classes and the predictor variables.

Using trees is a common method to visualise knowledge. From the biblical tree of knowledge to family histories, and Darwin's famous tree of life, decision trees are the latest generation of representing knowledge in a branched structure (Lima, 2014). Decision trees are grown with complex statistics, but they are a natural tool to make sense of the world. For example, the outcome of a decision tree to predict outdoor water consumption is visualised in Figure 13.5.

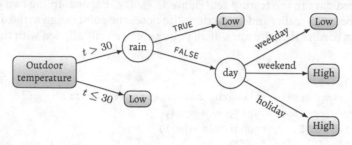

FIGURE 13.5 Decision tree example.

You can read this tree like a flow chart, using the following decision rules:

- Temperature > 30°C

    - Rain: Consumption is *low*

    - No rain

        * Weekday: Consumption is *low*
        * Weekend or holiday: Consumption is *high*

    - Temperature ≤ 30°C: Consumption is *low*

Several methods exist to grow decision trees, but they all follow the same principles. The first step determines the root node by determining which variable provides the best-defined split between outcomes. In our simple example above, more days with high water consumption occur when the temperature is over 30°C than below this threshold, so the algorithm uses this as the first node. The algorithm continues to the next level and finds the next split by determining which variable best predicts the outcome. These steps repeat until all splits perfectly predict the outcome. The algorithm also stops when a stopping criterion, such as a maximum depth, has been reached. Finally, the algorithm needs a decision rule to determine the optimal spile at each point. Several methods are available, such as Gini Impurity, Entropy, and Information Gain, which are not further explained in this chapter.

Decision trees are easy to interpret, as they can be intuitively understood without mathematical knowledge. The tree can predict outcomes for new variables by following the path from the root node to one of the leaf nodes. In the example above, we can predict that water consumption will be low when it is 32°C and there is no rain on a weekday. Decision trees are great in a non-linear context and can classify any type of variable. A disadvantage is that they can be inaccurate and unstable. Accuracy can be improved by combining many trees in a so-called Random Forest, but at the cost of interpretability. Decision trees have also been used to predict the compressive strength of concrete (Alghamdi, 2022), so we continue the concrete case study to demonstrate the principles of growing and evaluating decision trees.

13.5.1 Concrete Strength Case Study

Algorithmic trees exist in two forms, regression trees predict numeric outcomes, and classification trees develop rules to group data. In this example, we use a classification tree, so we first create categories of compressive strength. In engineering practice, concrete strength is specified in intervals of 10 MPa, so the first step in building this model is to cut the compressive_strength variable. The cut() function splits a vector into equal parts and assigns a label to each mixture. To reduce the number of permutations, we use classes with a span of 20 MPa. The compressive strength variable does not form part of the model because it perfectly correlates with the concrete class. The data sets are partitioned the same way as with regression models, but we create a new sample to prevent bias.

```
classes <- seq(0, 90, by = 20) # Create cutting points
concrete$class <- cut(round(concrete$compressive_strength),
                      breaks = classes,
                      labels = paste0("C", classes[-length(classes)] + 10))
concrete_class <- dplyr::select(concrete, -compressive_strength)

set.seed(123)
n <- nrow(concrete_class)
train_id <- sample(1:n, 0.7 * n)

concrete_class_train <- concrete_class[train_id, ]
concrete_class_test <- concrete_class[-train_id, ]

t <- nrow(concrete_class_test)
test_id <- sample(1:t, 0.5 * t)

concrete_class_test1 <- concrete_class_test[test_id, ]
concrete_class_test2 <- concrete_class_test[-test_id, ]
```

The *rpart* library provides an implementation of the CART (Classification and Regression Trees) algorithm to grow trees (Therneau & Atkinson, 2022). Running this algorithm follows the same principles as with regression analysis. The rpart() function needs a formula and a data set. The method = "class" parameter specifies that this is a classification model for categorical variables. For the concrete case study, we could have used a numeric model with method = "anova", but the purpose of this chapter is to demonstrate classification. The code below generates two models using the same principles as in the linear regression example.

```
library(rpart)

tree_min <- rpart(class ~ I(water / cement) +
                    coarse_aggregate +
                    fine_aggregate +
                    age_day,
                  data = concrete_class_train, method = "class")

tree_all <- rpart(class ~ ., data = concrete_class_train,
                    method = "class")
```

The output of the modelling variables shows the decision rules and their associated statistics.

Various parameters are available to tune the model, which are added to the rpart() function call. For example: rpart(class ~ ., data = concrete_class_train, method = "class", maxdepth = 1) creates a tree with only one decision point.

- minsplit: specifies the smallest number of observations in the parent node that could be split further. If you have less than minsplit records in a parent node, it is labelled as a terminal node (default is 20).

- `minbucket`: defines the smallest number of observations allowed in a terminal node (default: `round(minsplit / 3)`).

- `maxdepth`: restricts the depth of the tree (default: 30).

The *rpart* package can visualise a tree, but the *rpart.plot* package does a better job (Milborrow, 2022). The package has a lot of options to change the visualisation. The `extra = 2` parameter changes the information displayed in the boxes. The help file for this function contains all the details about fine-tuning the visualisation.

```
library(rpart.plot)
tree_vis <- rpart(class ~ ., data = concrete_class_train,
                  method = "class", maxdepth = 3)
rpart.plot(tree_vis, extra = 2)
```

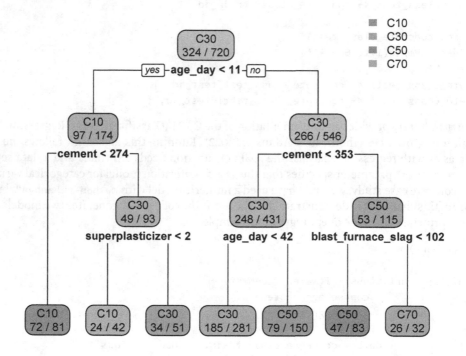

FIGURE 13.6 Visualising a decision tree.

The tree in Figure 13.6 uses a simplified model with a maximum depth of three steps. The information in the boxes shows the concrete class and how often it appears in the data applicable to that step. So in the first step, most samples are in class C30. The data is then split on curing age at 11 days. The next split uses the amount of cement, with C10 and C30 as the most common classes.

By convention, nodes to the left confirm the condition and to the right reject the condition. So to create concrete class C70, we need more than 11 days of curing, more than 353 kg cement, and less than 102 kg furnace slag. Note that this simplified model will most likely have low accuracy in the real world.

13.5.2   Cross-Validation for Classification Models

The last step in this decision tree example uses the testing data to validate the models. The `predict()` function works the same as when running linear regressions, so the code should look familiar. The prediction function results in a vector with modelled concrete classifications, which we can compare with the measured values in the respective testing sets.

```
predict_tree_min <- predict(tree_min,
                        newdata = concrete_class_test1, type = "class")

predict_tree_all <- predict(tree_all,
                        newdata = concrete_class_test2, type = "class")
```

For classification models, the common method to compare test predictions with observed data is a so-called confusion matrix. The rows indicate the actual values, and the columns indicate the predicted values. This means that the numbers in the diagonal are accurate predictions where the measured and predicted values match. The confusion matrix is a numeric expression of the graphs in Figure 13.2. The name does not stem from the fact that these tables can indeed be confusing to interpret but that they summarise confusion between prediction and observation.

- The values in the diagonal are the True Positives (TP), which for C10 is 22 and for C30 is 58.

- The False Positive (FN) is the sum of the relevant column, except for the True Positive, which for C10 is 8 and for C30 it is $4 + 5 + 0 = 9$.

- The False Negative (FP) is the sum of the relevant row, except for the True Positive. For C10, the False Positive is 4 and for C30 it is $8 + 10 + 1 = 19$.

- The True Negative (TN) is the sum of all values not associated with the relevant class, which is $58 + 10 + 1 + 5 + 27 + 2 + 0 + 4 + 13 = 120$ for C10.

```
confusion_matrix_min <- table(predict_tree_min, concrete_class_test1$class)
confusion_matrix_all <- table(predict_tree_all, concrete_class_test2$class)

confusion_matrix_all


predict_tree_all C10 C30 C50 C70
            C10  22   4   0   0
            C30   8  58  10   1
            C50   0   5  27   2
            C70   0   0   4  13

cm <- confusion_matrix_all

(tp <- diag(cm))
(fn <- colSums(cm) - diag(cm))
(fp <- rowSums(cm) - diag(cm))
(tn <- sum(cm) - (colSums(cm) + rowSums(cm) - diag(cm)))
```

```
C10 C30 C50 C70
 22  58  27  13
C10 C30 C50 C70
  8   9  14   3
C10 C30 C50 C70
  4  19   7   4
C10 C30 C50 C70
120  68 106 134
```

The overall accuracy of a prediction is defined by the total number of accurate predictions (sum of the True Positives) divided by the total number of predictions (sum of all values in the matrix).

```
sum(diag(confusion_matrix_min) / sum(confusion_matrix_min))
sum(diag(confusion_matrix_all) / sum(confusion_matrix_all))
```

```
[1] 0.6428571
[1] 0.7792208
```

A confusion matrix has a lot of options to calculate the sensitivity, specificity, and other metrics. The *caret* package provides the confusionMatrix() function, which provides a detailed analysis of the prediction and for each class.

Sensitivity is the True Positive rate, which reports on the probability of a positive prediction. Conversely, specificity is the True Negative rate, which reports on the probability of a negative prediction.

```
(sensitivity <- tp / (tp + fn))
(specificity <- tn / (tn + fp))
```

```
      C10       C30       C50       C70
0.7333333 0.8656716 0.6585366 0.8125000
      C10       C30       C50       C70
0.9677419 0.7816092 0.9380531 0.9710145
```

```
caret::confusionMatrix(predict_tree_all, concrete_class_test2$class)
```

```
Confusion Matrix and Statistics

          Reference
Prediction C10 C30 C50 C70
      C10  22   4   0   0
      C30   8  58  10   1
      C50   0   5  27   2
      C70   0   0   4  13


Overall Statistics

               Accuracy : 0.7792
                 95% CI : (0.7054, 0.842)
    No Information Rate : 0.4351
    P-Value [Acc > NIR] : < 2.2e-16

                  Kappa : 0.675
```

```
Mcnemar's Test P-Value : NA
```

```
Statistics by Class:
```

	Class: C10	Class: C30	Class: C50	Class: C70
Sensitivity	0.7333	0.8657	0.6585	0.81250
Specificity	0.9677	0.7816	0.9381	0.97101
Pos Pred Value	0.8462	0.7532	0.7941	0.76471
Neg Pred Value	0.9375	0.8831	0.8833	0.97810
Prevalence	0.1948	0.4351	0.2662	0.10390
Detection Rate	0.1429	0.3766	0.1753	0.08442
Detection Prevalence	0.1688	0.5000	0.2208	0.11039
Balanced Accuracy	0.8505	0.8236	0.7983	0.89176

In conclusion, just as was the case with the regression model, the decision tree with all parameters outperforms the parsimonious model. The decision trees are overall less accurate than a linear regression.

13.6 Further Study

This chapter barely scratches the surface on the topic of machine learning. The R language has access to many packages that implement machine learning methods. This richness of methods is a great advantage, but the problem is that each package has a different syntax and approach.

Max Kuhn developed the Caret (Classification and Regression Training) package, not to be confused with a carrot, to standardise machine learning (Kuhn, 2022). This package has functions to implement cross-validation, train, and tune the models and test predictions, as we briefly saw above.

While the Caret package is still actively maintained, Kuhn has developed a new set of packages following the Tidyverse way of analysing data, called Tidymodels (Kuhn & Wickham, 2020). This collection of packages is, at the time of writing, still in active development.

Regression trees are generally prone to overfitting, which will result in a lower accuracy when applying them to new data. An often used method to compensate for the tendency for decision trees to overfit is to build a Random Forest. This is a model that combines hundreds or even thousands of decision trees and uses statistics to optimise the model. While this method can achieve a high level of accuracy, Random Forests are impossible to interpret by humans due to their complexity.

14

In Closing

The Introduction to this book mentions that within the data science community, professionals who possess computer science and statistical skills, combined with expertise in the subject matter under analysis, are so rare they are known as unicorns. My primary motivation for writing this book is to introduce domain experts in water management to computer science through the R language. Data science unicorns can be created because we can teach subject matter experts computer science on top of the mathematical skills they already possess.

Core data science competencies are essential to leveraging digital capabilities in a quickly transforming industry. The best path toward becoming a digital water utility is not to rely on external expertise and purchase the latest gadgets but to educate established professionals. You might not write code as your day job, but understanding how it works will help professionals communicate with data scientists.

I hope this book has inspired you to learn the R language or any other data science language. This book never pretended to be a complete course in data science but a longitudinal cross-section of the capabilities of the R langue. This book also highlights that writing code is about more than just knowing the syntax and which function to use to create a result. For data science to be integrated with business operations, it must be reproducible and follow a clear workflow from problem statement to data product. Being a data scientist is not only about learning how to code but also how to leverage the workflow to create useful data products.

The most challenging part of writing a book like this is deciding what not to include. These chapters barely touch the surface of what can be achieved by writing data science code. The R language has many more capabilities than what was covered in these pages, such as spatial analysis and tools to develop online data applications. The *Further Study* sections in each chapter alluded to further capabilities of the R language.

Two additional capabilities in data science that are of value to water management are spatial and textual data analysis.

Most water utility data has a spatial component. Almost every measurement or observation is taken at a defined location, either by name or coordinate. Spatial analysis capabilities in R provide extensive abilities to analyse and visualise spatial data. The *Leaflet* package is a user-friendly application to publish interactive maps in R and visualise the results of your analysis (Cheng et al., 2022). More specialised packages can statistically analyse spatial data, produce accurate maps in various coordinate projections, and process shape files.

Although most water management data consists of time series of physical measurements, a customer-focussed water utility also collects textual data from its customers. Various libraries, such as Tidytext (Silge & Robinson, 2016), can analyse written text, such as customer feedback. A common approach is to use sentiment analysis to ascertain whether a message is positive or negative. This technique is prevalent in analysing social media posts.

The R language is truly a Swiss Army Chainsaw for data analysis. However, this book is only the first step in your journey to becoming a water data scientist.

14.1   Start Your Journey to Data Science

Now that you have completed this book, you are only at the start of your journey into data science. Your journey to mastery of data science with R has seven steps:

1.   Understand the basics
2.   Create simple programs
3.   Practice
4.   Ask for help
5.   Build projects
6.   Help others

This book has covered the basics, and now it is up to you to develop simple programs and practice your skills.

14.1.1   How to Ask For Help

The help entries in R are only helpful to intermediate or advanced users. Many packages have vignettes, which provide a more comprehensive description of the functionality. The next step is to find an answer using your favourite search engine. You will quickly realise that there are no problems that have not already been experienced and solved by somebody else. The R language is open-source, and many analysts share their code on websites and forums.

Your search engine will likely divert you to one of the many online forums where developers help each other. Websites such as Stackoverflow and Reddit have active communities where you can ask questions about R. If you are on Twitter or Mastodon, the #RStats hashtag connects fellow data scientists.

Before you post anything on these websites, check to see whether your question has not already been answered in a slightly different form. The best way to ensure you receive a helpful answer is to be as specific as possible. Add an example of your code and include some data called a *Minimum Working Example* (MWE). An MWE enables the community members to replicate your problem and increases your chances of receiving an answer. For example, you would like to know how to convert a wide data frame to a long version. In this case, you could provide a specific example that shows the before and after situation.

```
df_wide <- data.frame(A = c(1, 2),
                      B = c(12, 34),
                      C = c(43, 76),
                      D = c(5, 12))

df_long <- data.frame(A = c(1, 2, 1, 2, 1, 2),
                      var = c("B", "B", "C", "C", "D", "D"),
                      val = c(12, 34, 43, 76, 5, 12))
```

Most closed forums have internal rules. Make sure you familiarise yourself with these rules to increase the likelihood of receiving a helpful answer.

Once you become proficient and start developing projects, the best way to give back to the community is to help others. Helping others is not only an act of altruism. It is also the best way to become a master at your craft.

14.1.2 Learning Other Languages

This book discusses the R language as the weapon of choice to analyse data, but other languages are equally capable. Many websites discuss which programming language is best for data analysis. These discussions are primarily fruitless, and pragmatic data scientists use whichever language helps them solve the problem. To become a fully-skilled data scientist, you will have to also learn Python and SQL and auxiliary languages such as Regex, some HTML and Markdown.

The RStudio software can process Python and SQL code, and with R Markdown files, you can merge various languages. What is certain is that data science is a specialised disciple that is not a fad but a step change in how water professionals will manage and analyse data. The water industry will benefit when its workforce fully embraces data science principles so that we can secure plentiful and safe drinking water long into the future.

Bibliography

Aggarwal, C. C. (2016). An introduction to outlier analysis. In *Outlier Analysis* (pp. 1–34). Springer International Publishing.

Alghamdi, S. J. (2022). Classifying High Strength Concrete Mix Design Methods Using Decision Trees. *Materials*, 15(5), 1950.

Anderson, C. (2015). *Creating a Data-Driven Organization: Practical Advice from the Trenches*. Sebastopol, CA: O'Reilly Media Inc, first edition.

Anscombe, F. (1973). Graphs in statistical analysis. *The American Statistician*, 27(1), 17–21.

Bååth, R. (2012). The state of naming conventions in R. *The R Journal*, 4(2), 74.

Bache, S. M. & Wickham, H. (2022). *magrittr: A Forward-Pipe Operator for R*. R package version 2.0.3.

Bae, I. & Ji, U. (2019). Outlier detection and smoothing process for water level data measured by ultrasonic sensor in stream flows. *Water*, 11(5), 951.

Bearden, W. (2011). *Handbook of Marketing Scales: Multi-Item Measures for Marketing and Consumer Behavior Research*. Thousand Oaks, California: SAGE.

Bernaards, C. A. & I.Jennrich, R. (2005). Gradient projection algorithms and software for arbitrary rotation criteria in factor analysis. *Educational and Psychological Measurement*, 65, 676–696.

Beygelzimer, A., Kakadet, S., Langford, J., Arya, S., Mount, D., & Li, S. (2022). *FNN: Fast Nearest Neighbor Search Algorithms and Applications*. R package version 1.1.3.1.

Box, G. E. P. (1979). Some problems of statistics and everyday life. *Journal of the American Statistical Association*, 74(365), 1–4.

Briers, D. R. (2016). *biotic: Calculation of Freshwater Biotic Indices*. R package version 0.1.2.

Broman, K. W. & Woo, K. H. (2018). Data organization in spreadsheets. *The American Statistician*, 72(1), 2–10.

Bruner, G. (2012). *Marketing Scales Handbook. A Compilation of Multi-Item Measures for Consumer Behavior & Advertising Research*. Fort Worth, TX: GCBII Productions.

Bryman, A. & Bell, E. (2011). Ethics in business research. In *Business Research Methods*. Oxford: Oxford University Press, 3rd edition.

Burkov, A. (2019). *The Hundred-Page Machine Learning Book*. LeanPub.

Burns, A., Vardy, S., Rowe, M., & Murchinson, R. (2011). *Victorian Water Customer: Water Services Needs & Values Summary Report*. Technical Report 1, Accenture, Melbourne.

Caffo, B., Peng, R., & Leek, J. T. (2018). *Executive Data Science. A Guide to Training and Managing the Best Data Scientists*. LeanPub.

Chaubey, A. (2020). *Practical Concrete Mix Design*. Taylor & Francis Ltd.

Cheng, J., Karambelkar, B., & Xie, Y. (2022). *leaflet: Create Interactive Web Maps with the JavaScript Leaflet Library*. R package version 2.1.1.

Cominola, A., Nguyen, K., Giuliani, M., Stewart, R., Maier, H., & Castelletti, A. (2019). Data mining to uncover heterogeneous water use behaviors from smart meter data. *Water Resources Research*, 55(11).

Cominola, A., Preiss, L., Thyer, M., Maier, H. R., Prevos, P., Stewart, R., & Castelletti, A. (2021). An assessment framework for classifying determinants of household water consumption and their priorities for research and practice. In *MODSIM2021, 24th International Congress on Modelling and Simulation*.

Cronbach, L. J. & Meehl, P. E. (1955). Construct validity in psychological tests. *Psychological Bulletin*, 52(4), 281–302.

Dancho, M. & Vaughan, D. (2023). *anomalize: Tidy Anomaly Detection*. R package version 0.2.3.

Davies, R., Locke, S., & D'Agostino McGowan, L. (2022). *datasauRus: Datasets from the Datasaurus Dozen*. R package version 0.1.6.

Davis, K. & Patterson, D. (2012). *Ethics of Big Data*. Sebastopol, CA: O'Reilly.

DeVellis, R. F. (2011). *Scale Development: Theory and Applications*. Applied Social Research Methods. SAGE Publications, 3rd edition.

Fitzgerald, M. (2012). *Introducing Regular Expressions*. Sebastopol, CA: O'Reilly.

Frické, M. (2009). The knowledge pyramid: A critique of the DIKW hierarchy. *Journal of Information Science*, 35(2), 131–142.

Goetz, M. K. (2016). Why surveys work and how they can disappoint. *Journal AWWA*, 108(5), 70–74.

Gohel, D. & Skintzos, P. (2023). *flextable: Functions for Tabular Reporting*. R package version 0.8.5.

Grolemund, G. & Wickham, H. (2011). Dates and times made easy with lubridate. *Journal of Statistical Software*, 40(3), 1–25.

Gurung, T. R., Stewart, R. A., Sharma, A. K., & Beal, C. D. (2014). Smart meters for enhanced water supply network modelling and infrastructure planning. *Resources, Conservation and Recycling*, 90, 34–50.

Hahsler, M., Piekenbrock, M., & Doran, D. (2019). dbscan: Fast density-based clustering with R. *Journal of Statistical Software*, 91(1), 1–30.

Harrower, M. & Brewer, C. A. (2003). Colorbrewer.org: An online tool for selecting colour schemes for maps. *The Cartographic Journal*, 40(1), 27–37.

Hyndman, R. J. & Fan, Y. (1996). Sample quantiles in statistical packages. *The American Statistician*, 50(4), 361–365.

Hyndman, R. J. & Khandakar, Y. (2008). Automatic time series forecasting: the forecast package for R. *Journal of Statistical Software*, 26(3), 1–22.

ISO 1438 (2017). *Hydrometry — Open Channel Flow Measurement Using Thin-Plate Weirs*. Technical report, International Organization for Standardization, Geneva, CH.

ISO 8601 (2019). *Date and time — Representations for information interchange — Part 1: Basic rules*. Technical report, International Organization for Standardization, Geneva, CH.

Jenkins, S. & Storey, M. (2012). What Customers Really Want: Findings of a Value Segmentation Study. In *OzWater '12* Sydney: Australian Water Association.

Kaplan, J. (2020). *fastDummies: Fast Creation of Dummy (Binary) Columns and Rows from Categorical Variables*. R package version 1.6.3.

Kassambara, A. (2022). *ggcorrplot: Visualization of a Correlation Matrix using ggplot2*. R package version 0.1.4.

Kassambara, A. & Mundt, F. (2020). *factoextra: Extract and Visualize the Results of Multivariate Data Analyses*. R package version 1.0.7.

Kay, P. (2009). *The World Color Survey*. Stanford, California: CSLI Publications.

Knuth, D. E. (1974). Computer programming as an art. *Communications of the ACM*, 17(12), 667–673.

Komsta, L. (2022). *outliers: Tests for Outliers*. R package version 0.15.

Komsta, L. & Novomestky, F. (2022). *moments: Moments, Cumulants, Skewness, Kurtosis and Related Tests*. R package version 0.14.1.

Kotler, P., Brown, L., Burton, S., Deans, K., & Armstrong, G. (2010). *Marketing*. Frenchs Forest, N.S.W: Pearson Australia.

Kuhn, M. (2022). *caret: Classification and Regression Training*. R package version 6.0-93.

Kuhn, M. & Wickham, H. (2020). *Tidymodels: a collection of packages for modeling and machine learning using tidyverse principles*.

La, K. & Vinot, K. (2008). Customer Segmentation in the Residential Sector. *Water*, 35(February), 75–79.

Leys, C., Ley, C., Klein, O., Bernard, P., & Licata, L. (2013). Detecting outliers: Do not use standard deviation around the mean, use absolute deviation around the median. *Journal of Experimental Social Psychology*, 49(4), 764–766.

Lima, M. (2014). *The Book of Trees*. Abrams & Chronicle Books.

Machlis, S. (2019). *Practical R for Mass Communication and Journalism*. Chapman & Hall/CRC the R Series. Boca Raton, FL: CRC Press.

Maechler, M. (2023). *Rmpfr: R MPFR - Multiple Precision Floating-Point Reliable*. R package version 0.9-1.

Matejka, J. & Fitzmaurice, G. (2017). Same stats, different graphs: Generating datasets with varied appearance and identical statistics through simulated annealing. In *Proceedings of the 2017 CHI Conference on Human Factors in Computing Systems - CHI '17* (pp. 1290–1294). Denver, Colorado, USA: ACM Press.

McBride, G. B. (2005). *Using Statistical Methods for Water Quality Management: Issues, Problems, and Solutions*. Wiley Series in Statistics in Practice. Hoboken, NJ: Wiley-Interscience.

McCallum, Q. E. (2013). *Bad Data Handbook: Mapping the World of Data Problems*. Beijing: O'Reilly, first edition.

Meyer, D., Dimitriadou, E., Hornik, K., Weingessel, A., & Leisch, F. (2023). *e1071: Misc Functions of the Department of Statistics, Probability Theory Group (Formerly: E1071), TU Wien*. R package version 1.7-13.

Milborrow, S. (2022). *rpart.plot: Plot rpart Models: An Enhanced Version of plot.rpart*. R package version 3.1.1.

Monks, I., Stewart, R. A., Sahin, O., & Keller, R. J. (2021). Taxonomy and model for valuing the contribution of digital water meters to sustainability objectives. *Journal of Environmental Management*, 293(nil), 112846.

Murphy, R., Perry, E., Keisman, J., Harcum, J., & Leppo, E. W. (2022). *baytrends: Long Term Water Quality Trend Analysis*. R package version 2.0.8.

Müller, K. & Wickham, H. (2022). *tibble: Simple Data Frames*. R package version 3.1.8.

Neuwirth, E. (2022). *RColorBrewer: ColorBrewer Palettes*. R package version 1.1-3.

Parr, J. (2005). Local water diversely known: Walkerton Ontario, 2000 and after. *Environment and Planning D: Society and Space*, 23(2), 251–271.

Peng, R. (2011). Reproducible research in computational science. *Science*, 334(6060), 1226–1227.

Prevos, P. (2017). *Customer Experience Management for Water Utilities: Marketing Urban Water Supply*. London: International Water Association.

Prevos, P. (2019). *Principles of Strategic Data Science: Creating Value from Data, Big and Small*. Packt Publishing.

Puget, S., Beno, N., Chabanet, C., Guichard, E., & Thomas-Danguin, T. (2010). Tap water consumers differ from non-consumers in chlorine flavor acceptability but not sensitivity. *Water Research*, 44(3), 956–964.

Revelle, W. (2022). *psych: Procedures for Psychological, Psychometric, and Personality Research*. R package version 2.2.9.

Ryan, J. A. & Ulrich, J. M. (2022). *xts: eXtensible Time Series*. R package version 0.12.2.

Shron, M. (2014). *Thinking with Data: [How to Turn Information into Insights]*. Sebastopol, California: O'Reilly, first edition.

Sievert, C. (2020). *Interactive Web-Based Data Visualization with R, Plotly, and Shiny*. Taylor & Francis Ltd.

Sievert, C., Parmer, C., Hocking, T., Chamberlain, S., Ram, K., Corvellec, M., & Despouy, P. (2022). *plotly: Create Interactive Web Graphics via plotly.js*. R package version 4.10.1.

Silge, J. & Robinson, D. (2016). tidytext: Text mining and analysis using tidy data principles in r. *JOSS*, 1(3).

The Turing Way Community (2021). *The Turing Way: A Handbook for Reproducible, Ethical and Collaborative Research*. Zenodo.

Therneau, T. & Atkinson, B. (2022). *rpart: Recursive Partitioning and Regression Trees*. R package version 4.1.19.

Tierney, N. (2017). visdat: Visualising whole data frames. *JOSS*, 2(16), 355.

Torres-Matallana, J. (2021). *EmiStatR: Emissions and Statistics in R for Wastewater and Pollutants in Combined Sewer Systems*. R package version 1.2.3.0.

Tufte, E. (2001). *The Visual Display of Quantitative Information*. Graphics Press, second edition edition.

Tukey, J. W. (1977). *Exploratory Data Analysis*. Addison-Wesley.

Wexler, S., Shaffer, J., & Cotgreave, A. (2017). Data Visualization: A Primer. In *The Big Book of Dashboards* (pp. 1–36). Hoboken, NJ, USA: John Wiley & Sons, Inc.

Wickham, H. (2014). Tidy data. *Journal of Statistical Software*, 59(10), 1–23.

Wickham, H. (2016). *R for data science: import, tidy, transform, visualize, and model data*. Sebastopol, CA: O'Reilly Media.

Wickham, H. (2022a). *forcats: Tools for Working with Categorical Variables (Factors)*. R package version 0.5.2.

Wickham, H. (2022b). *stringr: Simple, Consistent Wrappers for Common String Operations*. R package version 1.5.0.

Wickham, H. (2022c). *tidyverse: Easily Install and Load the Tidyverse*. R package version 1.3.2.

Wickham, H. & Bryan, J. (2022). *readxl: Read Excel Files*. R package version 1.4.1.

Wickham, H., Chang, W., Henry, L., Pedersen, T. L., Takahashi, K., Wilke, C., Woo, K., Yutani, H., & Dunnington, D. (2022a). *ggplot2: Create Elegant Data Visualisations Using the Grammar of Graphics*. R package version 3.4.0.

Wickham, H., François, R., Henry, L., & Müller, K. (2022b). *dplyr: A Grammar of Data Manipulation*. R package version 1.0.10.

Wickham, H. & Henry, L. (2023). *purrr: Functional Programming Tools*. R package version 1.0.1.

Wickham, H., Hester, J., & Bryan, J. (2022c). *readr: Read Rectangular Text Data*. R package version 2.1.3.

Wickham, H., Vaughan, D., & Girlich, M. (2023). *tidyr: Tidy Messy Data*. R package version 1.3.0.

Wilkinson, L. (2005). *The Grammar of Graphics*. New York: Springer.

Wong, D. M. (2010). *The Wall Street Journal Guide to Information Graphics*. New York; London: 1W.W. Norton.

Xie, Y. (2017). *Bookdown: Authoring Books and Technical Documents With R Markdown*. Boca Raton, FL: CRC Press, Taylor & Francis Group.

Xie, Y. (2018). *R Markdown: The Definitive Guide*. Boca Raton, FL: CRC Press, Taylor & Francis Group.

Yeh, I.-C. (1998). Modeling of Strength of High-Performance Concrete Using Artificial Neural Networks. *Cement and Concrete Research*, 28(12), 1797–1808.

Yong, A. G. & Pearce, S. (2013). A Beginner's Guide To Factor Analysis: Focusing on Exploratory Factor Analysis. *Tutorials in Quantitative Methods for Psychology*, 9(2), 79–94.

Zaichkowsky, J. L. (1994). The personal involvement inventory: Revision, and application to advertising. *Journal of Advertising*, 23(4), 59.

Index

Printed in the United States
by Baker & Taylor Publisher Services